Institution of Civil Engineers

International Conference on

Maintenance Dredging II

Proceedings of the International Conference on Maintenance Dredging, organised by the Institution of Civil Engineers and held in Bristol, UK, on 6-7 May 2004

On behalf of:

Supported by:

Organising committee

John Land, Director of Dredging Research Ltd
Anthony Bates, Anthony D Bates Partnership
Nicola Meakins, Posford Haskoning

Published for the Organising Committee by Thomas Telford Publishing, Thomas Telford Ltd, 1 Heron Quay, London E14 4JD. www.thomastelford.com

Distributors for Thomas Telford books are
USA: ASCE Press, 1801 Alexander Bell Drive, Reston, VA 20191-4400, USA
Japan: Maruzen Co. Ltd, Book Department, 3–10 Nihonbashi 2-chome, Chuo-ku, Tokyo 103
Australia: DA Books and Journals, 648 Whitehorse Road, Mitcham 3132, Victoria

First published 2005

A catalogue record for this book is available from the British Library
ISBN 0 7277 3288 9

© Institution of Civil Engineers 2005 unless otherwise stated

All rights, including translation, reserved. Except for fair copying, no part of this publication may be reproduced, stored in a retrieval system or transmitted in any form or by any means, electronic, mechanical, photocopying or otherwise, without the prior written permission of the Books Publisher, Thomas Telford Publishing, Thomas Telford Ltd, 1 Heron Quay, London E14 4JD.

This book is published on the understanding that the authors are solely responsible for the statements made and opinions expressed in it and that its publication does not necessarily imply that such statements and/or opinions are or reflect the views or opinions of the publishers or of the Institution of Civil Engineers.

Printed and bound in Great Britain by Lightning Source UK Limited

Chairman's preface

In 1987 a highly successful conference on maintenance dredging was organised by ICE, and co-sponsored by CEDA and PIANC. Much has changed in this area since then. The regulatory framework within which maintenance dredging and the disposal of dredged materials are undertaken has changed beyond recognition and will change further with the implementation of the Water Framework Directive and the proposed EU Environmental Liability Directive. Dredging methods have improved and are more accurate, efficient and environmentally friendly. Methods of treatment to minimise problems arising during transport and disposal of dredged material have been refined. Treatment is becoming relatively commonplace and cost-effective, especially for small inland maintenance works. Our understanding of the potential environmental impacts of maintenance works, and methods to improve their prediction, although still far from perfect, have developed significantly.

A conference to review and evaluate these developments was therefore overdue. This conference was held in Bristol on 6-7 May 2004. Invited speakers, all specialists in their respective fields, presented papers dealing with a broad range of topics to provide a comprehensive update on all aspects of maintenance dredging including regulation, dredging methods, dredged material treatment and beneficial use, contract management, and environmental impact assessment and mitigation.

All those who are involved in the planning and execution of maintenance dredging, both coastal and inland, will find these conference papers informative and rewarding. A key objective of the conference was to provide delegates with state-of-the-art guidance on what is necessary to comply with regulation and what is possible in terms of current dredging practice.

John Land
Chairman, CEDA UK Section

Contents

1. Regulation

EU environmental directives and their implications for maintenance dredging
J. Brooke — 1

Regulation of marine maintenance dredging and disposal; a UK perspective
M. Pendle — 11

Regulation of inland waterway dredging and disposal
N. A. Smith — 19

Improve your chance of approval – beneficial use of dredged materials
J. Brien — 29

2. The practicalities of doing it

Hydrodynamic and non-dredging solutions
P. Hesk — 39

Recent developments in trailing suction dredgers
S. Vandycke — 49

Mechanical dredgers for maintenance and contaminated sediments
S. Vlug — 59

Optimising maintenance dredging
P. de Wit — 65

3. Inland projects

Dredging of Barton Broad; problems faced and lessons learnt
T. Wakelin — 72

Dredging and disposal methods for small projects (marinas, canals, lakes and reservoirs)
S. Bamford — 82

4. Contract management

Forms of contract and methods of measurement
J. M. Greenhalgh — 85

Contract *v* in-house maintenance dredging
    P. Mitchell   95

5. Environmental aspects

Modelling sediment transport and sedimentation
    M. P. Dearnaley, J. V. Baugh and J. R. Spearman   105

Predicting environmental impact
    P. Whitehead   115

Planning for environmental protection
    N. Clay   128

Methods of mitigating environmental impact
    A. Bates   137

Treatment and beneficial re-use of contaminated sediments
    G. S. Pomphrey and G. Van Dessel   150

Practical and effective environmental monitoring with reference to maintenance dredging for the Port of Maputo, Mozambique
    S. Challinor   160

Discussion   171

EU environmental directives and their implications for maintenance dredging

Jan Brooke, Environmental Consultant, Peterborough, UK

Abstract
Many EU environmental directives can affect the way in which maintenance dredging is carried out. Environmental issues can - and often do - influence the programming of dredging activities, the methods used, and the disposal options available.

Twenty years ago, environmental protection and maintenance dredging were rarely mentioned in the same breath. Whilst 1979 saw the Council Directive 79/409/EEC on the conservation of wild birds (the 'Birds Directive'), the full implications of this Directive for maintenance dredging only really started to become apparent in the mid 1990s. Since the Birds Directive, however, many other EU environmental directives have entered into force, each of which may impact on maintenance dredging decision-making to a greater or lesser extent.

This paper focuses on the practical and anticipated implications of three, linked, EU environmental directives: the Habitats Directive, the Water Framework Directive, and the Environmental Liability Directive.

Introduction
European environmental directives must be transposed into law and implemented in Member States in accordance with the Commission's requirements. The consequence of failing to implement the Directive properly can be infraction proceedings taken by the EC against the Member State(s) in question. Partly as a result of this threat of possible legal proceedings, it is clear when considering the effects of the various environmental directives forthcoming from the EC in the last twenty years or so, that there has been a considerable improvement in the quality of both inland and coastal/estuarine waters, and a marked strengthening in protection for, *inter alia*, sites of nature conservation importance.

To illustrate this strengthening in environmental protection, this paper considers the implementation of the three aforementioned Directives with respect to maintenance dredging in coastal and estuarine waters in England and Wales.

EU Habitats Directive
Council Directive 79/409/EEC on the conservation of wild birds (the 'Birds Directive') and Directive 92/43/EEC on the conservation of natural habitats and wild flora and fauna (the 'Habitats Directive') introduced a new and much-strengthened system of protecting sites of nature conservation interest. In England and Wales, these Directives were transposed into law by the 1994 Conservation (Habitats &c.)

Regulations – the 'Habitats Regulations'. Overall responsibility for delivering the requirements of the Directive rests with the Department for Environment, Food and Rural Affairs (Defra), with day-to-day implementation the responsibility of English Nature and Countryside Council for Wales, the Government advisors in each country (nature conservation being a devolved responsibility).

In addition to the responsibilities placed on the nature conservation organisations, the Habitats Regulations also have implications for competent authorities (statutory bodies with licensing or permitting powers) and for relevant authorities (statutory bodies with responsibility for the day-to-day management of the site). Many English and Welsh port authorities, as statutory bodies with both consenting powers and day-to-day management responsibilities are therefore both competent and relevant authorities under the Habitats Regulations.

The objectives of the Birds and Habitats Directives, as reflected in the Regulations, can be summarised as flows:
- To protect, manage and control all species of naturally occurring wild birds; to preserve diversity of habitats and conserve habitat of certain rare or migratory species
- To conserve biodiversity; to maintain or restore natural habitats and wild species at favourable conservation status through both site and species protection

These objectives are intended to be achieved, in part, by the designation and management of European sites - Special Protection Areas (SPA) under the Birds Directive, and Special Areas of Conservation (SAC) under the Habitats Directive. Regulations 33-36 of the Habitats Regulations make special provisions for so-called 'European Marine Sites' - any land covered (continuously or intermittently) by tidal waters or any part of the sea. Via these Regulations, the relevant authorities may establish a (statutory) management scheme under which they exercise their functions .. so as to secure compliance with the Regulations. This approach acknowledges the need for co-ordinated management of ongoing 'operations and activities' within coastal and estuarine waters – an area where management control is otherwise somewhat fragmented.

Maintenance dredging under the Habitats Directive: how it used to be ...
Initially when the implications of the Habitats Regulations for maintenance dredging were considered, it seemed reasonable to treat maintenance dredging as an ongoing operation or activity. Certainly, at the time of preparation of English Nature's Marine SACs project (ABP, 1999) it was clearly anticipated that this would be the case. Several port and harbour authorities therefore participated with other relevant authorities in setting up Management Schemes. The intention of such schemes was to collate relevant data, to assess the impacts of ongoing operations and activities against the achievement of the nature conservation objectives for the site in question, to modify activities where this proved necessary, and hence to demonstrate compliance with the requirements of the Habitats Regulations.

Maintenance dredging under the Habitats Directive: how it is now ...
However, whilst this approach appeared to provide a practical, common-sense way forward, legal considerations soon intervened. Defra have been advised that

maintenance dredging should be treated as a 'plan or project' – a decision which has potentially significant implications for those responsible for waterways which depend on maintenance dredging for their continued safe and efficient navigation.

Under Regulations 48-49 of the Habitats Regulations, competent authorities must determine whether a 'plan or project' requiring consent or authorisation is likely to have a 'significant effect' on a European site, either alone or in-combination with other plans and projects. If it is concluded that the proposed project will have a likely significant effect, the competent authority must undertake an 'appropriate assessment' which, in turn, will need to identify whether there will be an adverse effect on the integrity of the European site. This assessment involves the following stages:

- Data collection and collation
- Impact assessment
- Consideration/identification of appropriate mitigation measures
- Conclusion as to whether there will be an adverse effect on the integrity of the site and, if so:
- Assessment of alternatives and, if it is demonstrated that there are no alternatives:
- Consideration of possible imperative reasons of overriding public interest and, if there are such reasons
- Provision of compensation for lost or damaged habitats

The port and harbour industry has a great deal of experience of the appropriate assessment process insofar as it has been applied to a number of new development schemes frequently including capital dredging projects. The experience of many such developers has been one of increased costs, delays, and often real uncertainty about the likely outcome of the application in question. Further, the appropriate assessment process has highlighted other important questions including the difficulties of determining 'significance'; of assessing 'in-combination' effects; and of dealing with dynamic change (especially the issue of 'conservation' vs. 'preservation'). Concerns about the lack of flexibility and prescriptiveness of the process have also been expressed.

Industry working group
Against this background, the decision to treat maintenance dredging as a plan or project potentially subject to appropriate assessment requirements has not been accepted by all in the sector. Nonetheless, a working group comprising both industry representatives and representatives of the government regulatory bodies has been established. The aim of this group is to devise a system which, while compliant with the requirements of the Habitats Regulations, will allow maintenance dredging to continue without the need for disproportionate work. The work of the group recognises that maintenance dredging was typically ongoing before the designation of the European sites (ie. that nature conservation interest and maintenance dredging previously co-existed), and highlights the associated difficulties of 'proving a negative'. In this context, demonstrating that maintenance dredging is not having an impact - given that the activity was already ongoing at the time the conservation value of the site was assessed and judged to be worthy of protection - is challenging to say the least. The work of the group also recognises the importance of maximising the

consent period under a regime where appropriate assessment may be necessary in order to avoid the need to undertake unnecessarily frequent assessments.

Under the protocol being developed by the working group, it is proposed that future dredging applications will be backed by a 'baseline document', setting out the characteristics of the environment within which the dredging is proposed. The preparation of some such baseline documents is currently underway. As part of this process, it is proposed that the activity of maintenance dredging needs to be recognised as being part of the existing regime. As such, appropriate assessments for other proposed developments will need to take maintenance dredging and its implications into account (for example in the assessment of possible in-combination effects).

Pilot projects using the developing protocol are taking place (involving the Humber, Truro, and Cowes) and it is proposed that a system of 'shadow applications' (ie. applications accompanied by documentation produced using the protocol) be used, possibly beginning later in 2004. It is intended to complete the pilot study within a year or so, but it is important to ensure the continuation of existing approvals in the interim period.

Possible practical implications
Whereas agreement on the protocol is clearly required, it is also prudent to consider the possible practical implications for maintenance dredging should it be determined that maintenance dredging is likely to have a significant effect on a European site. In addition to the time and cost implications associated with undertaking an appropriate assessment, experience with capital dredging suggests that tidal or seasonal constraints might be imposed on dredging activities, that restrictions on overflow might need to be considered, or that choice of dredging method might be constrained, depending on the characteristics of the particular situation.

In considering the possible impacts of maintenance dredging, it is especially important to recognise some key differences between capital and maintenance dredging activities, most notably with respect to the environmental characteristics that might be affected. In the case of capital dredging, 'new' modifications are typically being made, potentially impacting previously undisturbed substrates and their associated habitats. Conversely, maintenance dredging - by its very nature – affects previously disturbed habitats in a physical environment where any changes have already been made. Notwithstanding this, best practice methods can – and should - be used to minimise impacts. Maintenance dredging and disposal methods which retain the sediment in the system may be preferred to those which remove material. Dredging may have less impact on the species using certain habitats if it is carried out at a certain time of the year – for example in April rather than in January if disturbance of over-wintering birds is identified as a potentially significant issue, or in March rather than in July if certain species of juvenile fish are shown to be at risk. An appropriate assessment may help to identify any such measures which are not already in place.

Finally, as indicated above, maintenance dredging has frequently been carried out for many years, generally pre-dating the designation of European sites. It would seem reasonable, therefore, to anticipate that any such dredging which has been causing a significant effect on a European site would already have been identified, and hence that the likelihood of maintenance dredging causing an adverse effect on the integrity of such sites should be relatively low. However, it is also worth noting that if a plan or project is shown to adversely affect the integrity of the European site, and assuming there are no alternatives, it must be demonstrated that there are imperative reasons of over-riding public interest, and that compensation must be provided. All of these tests are rigorous and none is without its difficulties.

EU Water Framework Directive

Directive 2000/60/EC 'establishing a framework for Community action in the field of water policy' (the Water Framework Directive) has been transposed into law England and Wales through the Water Environment (Water Framework Directive) (England and Wales) Regulations 2003. There are similar but separate regulations for the two cross-border river basin districts with Scotland. As with the Birds and Habitats Directives, overall responsibility for implementation rests with Defra in England and Wales, but in the case of the Water Framework Directive, the Environment Agency is the competent authority.

As a relatively new piece of legislation, there are still many uncertainties as to how the Water Framework Directive (WFD) will be implemented in practice. For example, the role of port and harbour authorities as statutory authorities under the Directive is still far from clear. However, in other respects, implementation is well-advanced and it is apparent that there could be a number of potentially significant implications (both benefits and challenges) for those involved in or responsible for maintenance dredging activities (see, for example, Brooke, 2003).

The purpose of the WFD is to update and consolidate existing piecemeal EU water legislation. It aims to establish a new, integrated approach to water protection, improvement and sustainable use. Its scope is extensive and its timescale is ambitious. As such, it has been widely described as the most important, far reaching water legislation ever to emerge from the EU.

An important difference between the WFD and other environmental legislation, however, is in its application. The WFD does not apply only to certain designated sites: it applies to all water bodies, including rivers, estuaries and coastal waters out to a minimum of one nautical mile. It also applies to man-made water bodies such as docks and canals, and to groundwater.

The following are among the key objectives of the WFD:
- to prevent deterioration in status
- to restore water bodies to good status by 2015, and
- to achieve the objectives for areas protected under other EC legislation

Chemical and ecological targets

Progress in achieving these objectives will be measured using both chemical and ecological criteria. However, in a manner which is once again different from other environmental directives, progress will be measured against targets which are typically derived not from the status quo or a recent condition, but from the unmodified ('pristine') state of the natural water body.

The unmodified state of the water body is referred to as high ecological status (HES). The target for surface water bodies will be good ecological status (GES), described as a 'slight reduction' in quality when compared to HES.

Some water bodies meeting certain strict criteria may be designated as 'heavily modified' (HMWB) or 'artificial' (AWB) water bodies. HMWBs are those where the existing physical modifications required for a particular use prevent the achievement of GES. However, for a water body to be designated as HMWB (or AWB), it must also be demonstrated that there are no technically viable, environmentally better options for achieving the aforementioned use which are not disproportionately expensive.

The target for HMWBs and AWBs will be good ecological potential (GEP), a slight reduction in quality when compared to the 'maximum ecological potential' (MEP) deemed to be achievable for that water body type (the latter to be derived from the closest natural water body type).

In addition to meeting GES/GEP by 2015, all surface water bodies will also have to achieve good chemical status, irrespective of whether or not they are identified as AWB or HMWB.

Derogations
The WFD also makes provision for less stringent targets to be set for any water body (HMWB, AWB, or natural) where it can be demonstrated that the achievement of the required quality will be disproportionately expensive or otherwise not feasible. Such temporary derogations must be set out in the river basin management plans (see below) and are subject to the approval of the Secretary of State.

A comprehensive monitoring programme is to be implemented by each Member State under the terms of the WFD, and this programme will be used *inter alia* to review progress in meeting the objectives and the continued validity (or otherwise) of derogations.

River Basin Management Plans
Among the other relevant requirements of the WFD is the preparation of statutory river basin management plans (RBMPs). These plans will describe the protection and restoration 'measures' necessary to ensure that the WFD environmental objectives are met. Measures may be statutory or non-statutory, national, regional or local. The RBMPs will provide the mechanism by which future water use and activities affecting the water will be managed.

Sediments
Discussion of the important role of sediments in the natural aquatic system is conspicuous by its absence in the WFD. Indeed, 'materials in suspension' are listed as a contaminant. Further, whereas coastal and transitional (estuarine) waters are discussed and provided for in the Directive, it is nevertheless apparent that there has been an initial focus in both WFD development and implementation, on fresh waters. There are also indications that land management techniques aimed at preventing sediment from entering water bodies are being discussed as possible measures, with seemingly no consideration for the possible implications of such an action for the

estuarine and coastal intertidal habitats which depend on land or river sourced sediments to facilitate their accretion.

In failing to recognise the important natural role of sediments, the WFD also misses an opportunity to properly differentiate between 'clean' and 'dirty' sediments. In respect of the latter, it is important to note that the WFD implements the 'polluter pays' principle, and should therefore provide the possibility for proper consideration of the question of responsibility for treating or disposing of contaminated sediments (ie. meeting the difference in costs between disposing of clean vs. contaminated materials). Related to this, the WFD should also offer the opportunity for improved source control with its consequent benefits for those involved in maintenance dredging.

Implications of the WFD for maintenance dredging
Maintenance dredging may thus be affected by the WFD, directly or indirectly, in a number of ways. Examples of possible implications include:
- additional temporal (tidal or seasonal) constraints on the dredging and disposal process;
- constraints on dredging method or on disposal options, etc.;
- reinstatement requirements;
- potentially prohibitive restrictions due to water status targets (eg. the need to achieve good chemical status);
- opportunities to determine responsibility for meeting the costs of treating or disposing of contaminated dredged materials;
- opportunities to achieve effective source control; and
- potential differences in interpretation and/or implementation of the WFD between Member States (the 'level playing field' issue).

In addition to the above, it is worth noting that future new development or modifications which affect water status will have to meet certain conditions under Article 4(7) of the WFD. The process for determining which developments are acceptable is similar in many respects to that discussed for 'plans and projects' under the 1994 Habitats Regulations (see above). Obviously, if new developments or modifications are constrained and/or delayed, any associated subsequent (capital and) maintenance dredging will similarly be affected.

Overall, therefore, it is important for those involved in maintenance dredging to become more aware of the WFD and to participate in its implementation. There are a number of important outstanding questions at both national and local level, ranging from the need to ensure consistent interpretation and implementation of the 'no deterioration' objective, to the role of the sector in the process of river basin management planning - and in particular determining the programme of measures.

EU Environmental Liability Directive
The EU Environmental Liability Directive was adopted in March 2004 and is due to enter into force in 2007. This Directive, the aim of which is to hold operators whose activities have caused environmental damage financially liable for the necessary remedial actions, enforces the 'polluter pays' principle. It applies to clearly identifiable polluters, shifting the cost of making good environmental damage to those who cause it. A key intention is that negligent operators will always be liable.

The Environmental Liability Directive introduces two distinct regimes, thus:
- professional operators conducting potentially risky activities will face strict liability even if they are not at fault. The Directive lists such activities in Annex III. Dredging and disposal are not specified, but related activities such as waste management and the release of pollutants into water are listed and it is not entirely clear whether or how dredging might be affected
- professional operators of all activities may be held liable (if they are at fault or negligent) for causing damage to species or natural habitats protected under the EU Birds and/or Habitats Directives.

An operator can be defined as 'a natural or legal public or private person who operates or controls the occupational activity': this includes a permit holder.

In addition to damage to biodiversity, the Directive also defines environmental damage to waters covered by the EU Water Framework Directive (all waters in the EU) and to land contamination where there is a risk to human health.

The Directive treats 'the environment' as a public good and regards public authorities as its guardians. Under the terms of the Directive, public bodies as competent authorities will have a duty to identify liable polluters and to ensure that they undertake or finance the necessary preventative or remedial measures. Should the competent authority fail to carry out this duty, the Directive makes provision for non-governmental organisations (or citizens affected by environmental damage) to require the competent authority to act.

The remedial measures required will depend on the nature of the damage. Competent authorities do have some discretion in this respect insofar as they must consider not only issues such as public health and safety but also benefits for the overall environment, the possibility of future and collateral damage, and social, economic and cultural concerns. Even so, the remedial measures have to be sufficient to make good or compensate adequately for the damage.

Limit on liability and the requirement for insurance
There is no financial limit on the amount that liable polluters will be required to pay to remedy environmental damage.

Operators are not currently required to take out financial security (indeed, there is currently a real lack of appropriate products). However, under the terms of the Directive, Member States must 'encourage the development of financial security instruments and markets'. There will be a review of the possible need for mandatory insurance cover in 2010.

Exemptions from liability
Exemptions from liability are limited. Environmental damage caused by *force majeure* (storms, armed conflicts, etc.) will not give rise to liability. Member states may exempt operators who can show that they were complying with the conditions set out in a licence or permit as long as the operator was not at fault or negligent. Member states may similarly exempt operators who can demonstrate that state of the

art methods were in use (ie. the activities were not considered likely to cause damage at the time they took place).

Diffuse pollution is not covered by the Environmental Liability Directive: it was considered that this would be ineffective and impractical, and that better controls potentially exist via other legal instruments.

Implications for maintenance dredging

Annex III of the Directive lists the occupational activities to which strict liability will be applied. Neither dredging nor disposal is specified. However, as indicated earlier, waste management and landfill operations subject to permit or registration under the waste, hazardous waste and/or landfill Directives are covered. So too are all discharges which require authorisation under, *inter alia*, the Water Framework Directive. The implications for maintenance dredging are therefore far from clear.

It is also important to remember that the Environmental Liability Directive applies to all significant damage to protected species and habitats if the operator is at fault or negligent, irrespective of the nature of the operation. Given the proportion of maintenance dredging that currently takes place in the vicinity of sites designated under the Birds and Habitats Directives, and the difficulties in determining what constitutes a 'significant' impact (see above), this is potentially of real concern.

A strategy for dealing with EU environmental directives

Recent years have seen environmental awareness accompanied by increasing numbers of EU environmental directives. Many of these have had and will continue to have significant implications for maintenance dredging. The three directives discussed in this paper, particularly when considered together, have some potentially far-reaching consequences for those responsible for and/or involved in dredging. It is clear that many operators and/or competent authorities need to:
- better understand the environment in which they operate;
- be not only aware of, but also up-to-date with, the legislation; and
- ensure that they are well-prepared to deal with its consequences.

Developing a good understanding depends on data. Data collection and management, dissemination and communication, planning and monitoring are all important. The development and implementation of a structured environmental management and decision-support framework, such as those developed for the Port of London Authority (see Clay, 2004) and for Gloucester Harbour Authority (personal communication, GHT, 2004), is likely to become of increasing value to many other port, harbour and navigation authorities. The use of such a structured procedure, based on a comprehensive data set, to inform decision-making can help to ensure that all relevant issues are identified and dealt with and – importantly – that nothing is left out.

Acknowledgements

The author is grateful to Dick Appleton, Harbour Engineer, Poole Harbour Commissioners and Chairman of the British Ports Association Environment Group for his advice on the development of the maintenance dredging protocol currently being progressed under the 1994 Habitats Regulations.

References

ABP, 1999. Good practice guidelines for ports and harbours operating within or near UK European marine sites. Natura 2000, UK Marine SACs Project. English Nature, Peterborough. July 1999.

Brooke, Jan. 2003. 'As Clear As Mud: The EU Water Framework Directive and its Possible Implications for Navigation Dredging'. Paper presented to the Central Dredging Association (CEDA) 'Dredging Days' Conference, Amsterdam, The Netherlands. 20th-21st November. janbrooke@compuserve.com

Clay, Nicola. 2004. Planning for environmental protection. Presented to CEDA Maintenance Dredging II, Bristol, 6-7 May 2004.

European Commission. 2004. Questions and Answers, Environmental Liability Directive. Press release Memo/04/78, Brussels, 1 April 2004. www.europa.eu.int

Regulation of Marine Maintenance Dredging and Disposal; a UK perspective.

Marie Pendle, Centre for Environment Fisheries and Aquaculture Science, UK

Introduction
Maintenance dredging and disposal constitutes one of the major anthropogenic perturbations to the marine environment, with considerable potential for contamination transfer. In the UK, 25 - 40 million wet tonnes are licensed for disposal each year, at a range of marine disposal sites. The House of Commons' Environment, Food and Rural Affairs Committee has reported that the current legislative and institutional framework governing this activity is fragmented and complex. This paper summarises the UK, European and international legislation relevant to dredging and disposal activities, highlighting the historic evolution and drivers. The pertinent legislation includes the 1972 London Convention(LC), its 1996 protocol (yet to be ratified by the required 26 countries), the 1992 OSPAR Convention, various EU Directives and the UK Food and Environment Protection Act 1985 (FEPA) and Coast Protection Act 1949 (CPA). Clear relationships exist between LC, OSPAR and FEPA, but the regulations controlling dredging and disposal of dredged material have been complicated by subsequent EU Directives. The EU Directives have resulted from disparate drivers and are not necessarily mutually compatible. The paper also seeks to elucidate the current decision-making process leading to the provision of consents and licences within England and Wales.

Regulations
Globally, ports and shipping still provide a lion's share of the transport of goods internationally. This means that dredging navigational routes and relocating the dredged material are of considerable economic and social interest. Historically, the sea and waterways have also been indiscriminately used for waste disposal. Occurrences such as Minamata, where many Japanese were poisoned by an industrial mercury effluent being released into coastal water, provided evidence that man's activities were degrading the marine environment. Since many nations relied on the seas to provide food and it was recognised that contamination and sea creatures do not respect territorial boundaries, the United Nations Conference on the Human Environment, held in Stockholm in 1972, provided an Action Plan for the Human Environment that included a recommendation for action at the international level on marine pollution. This political will resulted in the signing of the London Convention in 1972.

Convention on the Prevention of Marine Pollution by Dumping of Wastes and other Matter, 1972 (London Convention)

The drivers behind this international agreement were clearly reflected within the wording of Article 1 of the document as the following excerpt shows.

"... promote the effective control of all sources of pollution of the marine environment, and pledge themselves especially to take all practicable steps to prevent the pollution of the sea by the dumping of waste and other matter that is liable to create hazards to human health, to harm living resources and marine life, to damage amenities or to interfere with other legitimate uses of the sea."

There are economic drivers (the "living resources" and "legitimate users"), social drivers (the "hazards to human health" and "amenities") and environmental drivers (the "marine life").

The word "dumping" in the title was partially defined as "any deliberate disposal into the sea of wastes or other matter from vessels, aircraft, platforms or other man-made structures at sea". The practice of dredging material from ports, docks and navigational channels and taking it to other areas in the marine environment is covered by this agreement. There had been no previous international legislation which had controlled either dredging or dredging disposal, although there may have been individual national controls, either through legislation or voluntary measures.

Within the agreement, a list of priority hazardous substances were provided, either classified as Annex I or Annex II substances, see Table 1.

Annex I	Annex II

• Organohalogen compounds.
• Mercury and mercury compounds.
• Cadmium and cadmium compounds.
• Persistent plastics and other persistent synthetic materials, for example, netting and ropes, which may float or may remain in suspension in the sea in such a manner as to interfere materially with fishing, navigation or other legitimate uses of the sea.
• Crude oil and its wastes, refined petroleum products, petroleum, distillate residues, and any mixtures containing any of these, taken on board for the purpose of dumping.
• Radioactive wastes or other radioactive matter.
• Materials in whatever form (e.g. solids, liquids, semi-liquids, gases or in a living state) produced for biological and chemical warfare
	• Wastes containing significant amounts of the matters listed below:
	  arsenic)
	  beryllium)
	  chromium)
	  copper) and their compounds
	  lead)
	  nickel)
	  vanadium)
	  zinc)
	  organosilicon compounds
	  cyanides
	  fluorides
	  pesticides and their by-products not covered in Annex I.
	• Containers, scrap metal and other bulky wastes liable to sink to the sea bottom which may present a serious obstacle to fishing or navigation.

Table 1. Priority hazardous substances from the London Convention (excerpts from text)

The London Convention also provided the suggestion that signatory states, with "a common interest in particular geographic areas", should enter into supplementary agreements in order to improve the protection of the marine environment.

Convention for the Protection of the Marine Environment of the North-East Atlantic, 1992 (OSPAR Convention)

The OSPAR Convention is one of these supplementary agreements and is an agreement between Belgium, Denmark, Finland, France, Germany, Iceland, Ireland, Luxembourg, the Netherlands, Norway, Portugal, Spain, Sweden, Switzerland and the United Kingdom of Great Britain and Northern Ireland. The wording very much follows the pattern set out by the London Convention, but introduces the further concepts of ;
- ➤ the precautionary principle
- ➤ the polluter pays principle
- ➤ application of best available techniques and best environmental practice as so defined, including, where appropriate, clean technology.

OSPAR has facilitated harmonised processes, so that the reporting, measuring and monitoring required for the London Convention is in fact provided through the OSPAR route. The joint assessment of the quality status of the marine environment thus produced includes evaluations of how successful the measures taken have been, any measures planned and the identification of priorities for action.

It has also been made clear that the convention could not be used as a reason to prevent providing more stringent measures "with respect to the prevention and elimination of pollution... against the adverse effects of human activities" and a further article provided for public access rights to the data

Food and Environment Protection Act 1985 (FEPA)

Following the London Convention, the first UK legislation to transpose the international requirements to domestic law was the Dumping at Sea Act 1974, but this has long since been superseded by FEPA Part II. The wording again reflects its descent from the international conventions, requiring the authority to
"have regard to the need-
(i) to protect the marine environment, the living resources it supports and human health; and
(ii) to prevent interference with legitimate uses of the sea..."
whilst determining whether to issue a licence and including provisions within any licence necessary to achieve these aims.

Coast Protection Act 1949 (CPA)

It can be noted that all of the above regulations only deal with the disposal of dredged material. The United Kingdom has historically relied on its seafaring to provide trade, employment, food and defence. It has many ports which have been established for centuries and accompanied by dredging to provide safe navigation for ships arriving or leaving these ports. Reflecting this long history, the act of dredging an area has been regulated for many years, mainly through the Coast Protection Act 1949. There are exemptions to this, as there are individual arrangements under the

Harbours Act 1964, which appoint Port Authorities to consent the act of dredging in areas defined by their particular Harbour Order. These arrangements reflect the original concern attached to dredging, which was its potential for interference with navigation.

EU Directives
There are many different EU Directives that have some influence over the regulation of anthropogenic activities in the marine environment. The difficulty in implementation of all of them is that they arise from very different concerns, are transposed by adding to existing legislation and administered by different government departments and agencies. Below is a list of some of the main directives that can affect the regulation of dredging and dredged material disposal
- Pollution caused by certain dangerous substances discharged into the aquatic environment
- Quality required of shellfish waters
- Suitability of shellfish for human consumption
- Quality of bathing waters
- Conservation of natural habitats and of wild fauna and flora
- Conservation of wild birds
- A framework for community action in the field of water policy
- Assessment of the effects of certain plans and programmes on the environment

The first, Dangerous Substances in Water Directive links quite neatly with the hazardous substances provisions of the London Convention and continues the drive towards cleaner rivers and seas. The Shellfish Waters Directive requires that water quality is suitable for the protection of shellfish, so that the shellfish can grow and reproduce, protecting harvesting areas as a living resource. The Shellfish Hygiene and Bathing Waters Directives deal with microbiological standards, both with the aim of protecting human health. The Habitats and Birds Directives are both providers of biodiversity protection, seeking to conserve the marine environment. The Water Framework Directive seeks to overarch many of these directives to provide a holistic driver to safeguard the aquatic environment. The Environmental Impact Assessment Directive seeks to ensure that the ramifications of any sizeable project are thoroughly considered during the planning process.

United Kingdom
The two main pieces of legislation, FEPA 1985 and CPA 1949, seek to meet the responsibilities of the EU Directives, despite originally being imposed for quite different reasons. Until recently, licences and consents issued through this regulatory framework were assessed and awarded by different government departments. At present, within the UK, the regulatory authorities are defined geographically. For FEPA, the authority for Scotland is Fisheries Research Services (FRS), for Northern Ireland is Environmental and Heritage Services (EHS) and for England & Wales is Department for Environment, Food and Rural Affairs (Defra). Currently, in England and Wales, Defra has teamed up with the Department for Transport (DfT) to provide a single point of contact to apply for both FEPA licence and CPA consent. This single point is the Marine Consents Environment Unit (MCEU). The decision-making process behind the issue is now considered.

The decision-making process is highly complex, with multiple advisors each following different remits, as well as other stakeholders requiring consultation. MCEU sends out copies of the applications to all of the consultees determined to be appropriate for the application under consideration. At the barest minimum for FEPA, this will include the Defra Sea Fisheries Inspectorate (SFI) office covering the appropriate area and a Regulatory Assessment Team member from the Centre for Environment, Fisheries and Aquaculture Science (CEFAS), although this is generally only the case for existing licences which require renewal. The SFI provides advice regarding the impact on local fishing activities, based on a sound knowledge of their geographic area. CEFAS considers a range of impacts and will liase with the applicant to ensure that the material to be disposed has had the contamination assessed at CEFAS Laboratories. Under OSPAR guidelines, samples representing the material to be dredged must be provided at least every three years. The minimum number of sample stations is defined by the guidelines and in areas where there are contamination concerns, the material may require sampling to be undertaken on a more frequent basis, likewise, if the conditions surrounding the material to be disposed changes, further sampling may also be undertaken. CEFAS provides analysis of the material to a standardised quality, with high resolution and thoroughly tested and quality assured methodology. Another advantage to the sample analysis being provided by a single independent laboratory is that this enables Defra to ensure that the results can be part of an overall quality assessment of the sediment contaminant loadings around the coastline of England and Wales. The time involved in requesting samples, waiting for receipt and then analyses, provides one of the major constraints of the process of providing advice to Defra for consideration prior to the issue of a FEPA licence. The information, provided by the sample analyses results, is made available through a Public Register (results can also be supplied to applicants if specifically requested) and the mean contaminant levels in the material are considered. In order to assess the suitability of the material for disposal to sea, there are a number of factors addressed in CEFAS' consideration and in their advice to Defra. These include:

➢ Dredged area, material type and quantity
  CEFAS consider how regularly the area has been dredged previously, determining if the operation is ongoing, episodic or periodic. The physical characteristics of the material are described along with the overall size of the project.
➢ Dredge material quality
  The contaminant loading of the material is assessed against the geologic background loading of the material and "action levels", which have been determined according to scientific understanding of the environmental effects of the contaminant. It is important to point out that the action levels do not provide a pass/fail criteria, but are used as part of a 'weight of evidence' approach. These values will be used in conjunction with a range of other assessment methods, e.g. bioassays, as well as historical data and knowledge regarding the site. This integrated approach is in line with recent discussions regarding 'weight of evidence' approaches to environmental management of sediments. It considers balancing multiple lines of evidence concerning ecological assessment as an aid to decision making.
➢ Dredging Method
  The method to be used may change the physical characteristics of the material, or alter the potential for resuspension of contaminants into water, therefore its suitability is considered.

➢ **Alternatives to sea disposal**
  There is an obligation that alternatives should be considered prior to choosing the sea disposal option, so this is considered.
➢ **Disposal site**
  CEFAS will consider whether the disposal site has suitable characteristics for the material to be disposed and the history of disposal use, taking into account the impacts both at the disposal site itself and the surrounding environment, including fishery resources, biodiversity, physical changes, sediment transport and other factors.
➢ **Disposal contractor and vessels**
  Similarly to the dredging method to be used, the vessel to be used to discharge the material may alter the material's characteristics, so consideration of the vessel is included.
➢ **Licence Conditions**
  If any possible impacts can be mitigated by the inclusion of licence conditions, these are suggested for consideration by the MCEU.

There are normally more consultees, commonly these include;
➢ the Environment Agency, who will comment on water quality issues, flood defence issues and conservation issues
➢ English Nature, who will comment on any possible impacts on adjacent conservation areas,
➢ the Maritime and Coastguard Agency, who will comment on any navigational issues
➢ representatives of leisure users such as the Royal Yachting Association or the local Angling Association, who will comment on interference with leisure activities
➢ Sea Fishery Committees, who will comment on commercial fishery issues

This list is by no means exhaustive, but provides a flavour of the type of consultation that MCEU carries out. The final decision on the issue of a FEPA licence and the conditions contained within it rests with the MCEU. Similar consultation takes place for the CPA consent and includes consideration of whether an Environmental Statement or Appropriate Assessment for the project are required under the Harbour Works (Environmental Impact Assessment) Regulation 1999 and the Conservation (Natural Habitats etc) Regulations 1994 respectively, including scoping advice if either are necessary.

Occasionally the consultees will provide different perspectives and this can lead to protracted discussions to resolve the differences. If applicants submit scoping documents for their project to MCEU prior to the application, many of these discussions can take place first and the process can be more targeted and the information requirements of each of the consultees established early. This should streamline the process as far as possible.

The future of regulation
Future marine environmental regulation will be influenced by factors such as;
- the Water Framework Directive (WFD), as an umbrella over existing water EU Directives
- the Strategic Environmental Assessment Directive, allowing reference to a co-ordinated planning framework
- the suggestion of a new Marine Act, taking account of EU Directives as well as the London and OSPAR Conventions.

Concerns have been expressed that meeting the requirements of the Water Framework Directive will be prohibitively expensive, however, the shift from the principle of individual directives to an overarching directive is an opportunity to influence future regulation of marine maintenance dredging and sea disposal of dredged material and move towards a more coherent approach. The involvement of the Port, Harbour and Dredging Industries in the development of the UK and European approaches to implementing the WFD is important to ensure that the socio-economic aspects are understood alongside the environmental ones.

Currently, many small operators, such as marinas, need to undertake individual environmental impact assessments prior to carrying out their dredging campaigns. These can be costly even if tightly focussed and many of these operators are at the mercy of the consultants that they employ. The Strategic Environmental Assessment Directive will require that the regulators provide a document considering the vulnerabilities of a specific area, allowing individual project environmental impact assessments to focus their investigations, particularly with regard to cumulative impacts.

There have also been a number of recent reviews on aspects of regulation of human activities in the marine environment, including a call to consider the revision of the entire marine consents process in the United Kingdom. The possibility of provision of new legislation in the form of a 'Marine Act' is in the process of being considered. However, the provision of new legislation is extremely time consuming, therefore, a comprehensive Act is unlikely in the short term. The dredging industry and port, harbour and marina operators are encouraged to engage with the regulatory authorities and enable a working partnership to provide a more efficient, transparent approach to the sustainability of our coastline.

References

Chave, P. 2001. The EU Water Framework Directive: An Introduction. IWA Publishing, London.

Coastal Protection Act, 1949. London, HMSO.

Final Act of the Inter-governmental Conference on the Convention on the Dumping of Wastes at Sea, 1972. London, HMSO.

Food and Environmental Act, 1985. London, HMSO.

JAMP Guidelines for Monitoring Contaminants in Sediments

London Convention Scientific Group 2002. Guidelines for the Sampling of Sediment intended for Disposal at Sea. In: Report of the twenty fifth Meeting of the Scientific Group. 27th-31st May 2002. International Maritime Organisation.

Parker, M.M., 1987. The future for the disposal of dredged material in the UK. In: Maintenance Dredging: proceedings of a conference organized by the Institution of Civil Engineers and held in Bristol on 20-21 May, 1987. Thomas Telford, London.

1996 Protocol to the Convention on the Prevention of Marine Pollution by the Dumping of Wastes and other matter, 1972. International Maritime Organisation.

Oslo and Paris Commissions, 1993. Annex I Convention for the Protection of the Marine Environment of the North-East Atlantic. In: Ministerial Meeting of the Oslo and Paris Commissions, Paris 1992.

OSPAR Guidelines for the Management of Dredged Material

The Conservation (Natural Habitats etc) Regulations 1994. Statutory Instrument 1994 No. 2716. London, HMSO.

The Harbour Works (Environmental Impact Assessment) Regulations 1999. Statutory Instrument 1999 No. 3445. London, HMSO

Vivian, C.M.G., 2002. Regulation of Dredged Material Disposal at Sea in North-West Europe by the OSPAR Convention. In *'Dredging '02, Key Technologies for Global Prosperity'*, May 5-8 2002, Florida, USA, (Ed.) Stephen Garbiaciak, Jr., American Society of Civil Engineers.

Regulation of Inland Waterway Dredging and Disposal

N. A. Smith, British Waterways, Leeds, UK

Dredging is an important element in the sustainable management of canals and rivers. It provides for navigation and helps to maintain wildlife and fishery interest. Whilst dredging of inland waterways is primarily for navigation, some inland waterways are dredged for drainage and flood prevention. In some cases, dredging has been carried out primarily for environmental improvement. The need to dredge varies from waterway to waterway and is dependent on such things as sedimentation rates and usage. Generally the dredging cycles range from annual to 25 years. The quality of the sediment is also variable although the bulk of it tends to be inert or non-hazardous. Some lengths of the waterways contain sediments which have been contaminated by historical industrial activities. In a limited number of instances, the levels of contamination are significant and have been sufficient to classify the sediment as special/hazardous waste. The level of contamination within the sediments is generally reducing through regular maintenance dredging and the ongoing process of waterway restoration.

Canals have been in existence for some 200 years and for the majority of this time there were no legislative controls on the disposal of dredgings. During this period, dredgings were spread on banks and adjacent land and, in many instances, in dedicated disposal facilities (tips). When Regulations first appeared in 1988 there was no initial wholesale change in the disposal operations except for delays in starting projects due to the need to obtain consents and licenses.

Overview of the Legislation Affecting the Disposal of Dredgings

Since the 1988 Regulations there has been ongoing development of waste legislation, much of it influenced by EU Directives, which has become progressively more complex. Such is the current complexity, that the disposal of dredgings has become over regulated. This has had the consequence of increasing project lead in times, limiting the disposal options and increasing costs.

The following extract from the current list of waste legislation highlights the extent of the controls affecting the disposal of waste:

Extract of current waste legislation, May 2004

- *Control of Pollution (Amendment) Act 1989*
- *Environment Act 1995*
- *Environmental Protection Act 1990*
- *Controlled Waste (Registration of Carriers and Seizure of Vehicles) Regulations 1991, SI 1624*
- *Controlled Waste (Registration of Carriers and Seizure of Vehicles) (Amendment) Regulations 1998, SI 605*

Maintenance Dredging II, Thomas Telford, London, 2005.

Extract of current waste legislation, May 2004 (cont.)

- Controlled Waste Regulations 1992, SI 588
- Controlled Waste (Amendment) Regulations 1993, SI 566
- Environmental Protection (Duty of Care) Regulations 1991, SI 2839
- Landfill (England and Wales) Regulations 2002, SI 1559
- Special Waste Regulations 1996, SI 972
- Special Waste (Amendment) Regulations 1996, SI 2019
- Special Waste (Amendment) Regulations 1997, SI 251
- Special Waste (Amendment) (England and Wales) Regulations 2001 SI 3148
- Transfrontier Shipment of Waste Regulations 1994, SI 1137
- Waste Management Licences (Consultation and Compensation) Regulations 1999, SI 481
- Waste Management Licensing Regulations 1994, SI 1056
- Waste Management Licensing (Amendment) Regulations 1995, SI 288
- Waste Management Licensing (Amendment No 2) Regulations 1995, SI 1950
- Waste Management Licensing (Amendment) Regulations 1996, SI 1279
- Waste Management Licensing (Amendment) Regulations 1997, SI 2203
- Waste Management Licensing (Amendment) Regulations 1998, SI 606
- Waste Management Licensing (Amendment) (England) Regulations 2002, SI 674
- Waste Management Licensing (Amendment) (England) Regulations 2003, SI 595
- Waste Management Regulations 1996, SI 634

The Collection and Disposal of Waste Regulations 1988

These regulations were enacted under the Control of Pollution Act 1974 and came into force fully on 3 October 1988. Schedule 3 of the Regulations *(Waste to be treated as industrial waste)* included *"Waste from dredging operations"* and, for the first time, the disposal of dredgings became regulated. This was somewhat of a surprise for the navigation and dredging industry as it was not anticipated and it was soon discovered that the impact of the regulations would be significant.

These regulations required the disposal of dredgings to be regulated by a Waste Disposal License unless the operations qualified for an exemption under Schedule 6. The two exemptions which were most relevant to the disposal of dredgings are paragraphs 8 and 13.

Schedule 6, Para. 8 exemption of the Collection & Disposal of Waste Regs 1988

8. - (1) Subject to sub-paragraphs (2) and (3), the deposit of waste from dredging operations of any inland water within the meaning of section 135 of the Water Resources Act 1963.

(2) The deposit is made along the banks of the inland water from which the waste is dredged and is made as operations proceed. .

(3) The waste is not deposited in a lagoon or container.

Schedule 6, Para. 13 exemption of the Collection & Disposal of Waste Regs 1988

13. - (1) Subject to sub-paragraphs (2) and (3), the deposit -
  (a) of sewage sludge on land for the purpose of fertilising or otherwise beneficially conditioning that land; or
  (b) of any waste, on land used for agricultural purposes, for the purpose of fertilising or otherwise beneficially conditioning that land.

  (2) The waste is deposited directly onto the land and not in a lagoon or container.

> **Schedule 6, Para. 13 exemption of the Collection & Disposal of Waste Regs 1988**
>
> (3) The person depositing the waste shall furnish particulars to the disposal authority in whose area the deposit is to be made as follows -
> (a) Where there is to be a single deposit of waste, he shall furnish the following particulars in advance of making the deposit:
> (i) his name, telephone number and address;
> (ii) a description of the waste, including the process from which it arises;
> (iii) an estimate of the quantity of the waste; and
> (iv) the location and intended date of the deposit.
> (b) Where there are to be regular or frequent deposits of wastes of a similar composition he shall furnish the following particulars every six months:
> (i) his name, telephone number and address;
> (ii) a description of the waste, including the process from which it arises;
> (iii) an estimate of the total quantity of waste he intends to deposit during the next six months; and
> (iv) the locations and frequency of the deposits,
> and he may deposit wastes of a different description from that notified provided that he furnishes amended particulars in advance of making the deposit.

As a consequence of these Regulations, British Waterways set up a new team in 1990 specifically to obtain the necessary licenses and other consents required.

The Environmental Protection Act 1990

This Act provides the main statutory framework in relation to waste. In particular it:

- Defines waste
- Outlines the role and function of the waste regulation authorities (Environment Act 1995 created, and transferred these to, the new Environment Agencies).
- Established the criminal offences in relation to waste
- Lays down the waste management licensing system
- Established the statutory Duty of Care.

It is Part II of the Act which sets out the waste management and disposal requirements that affect those producing controlled waste as defined in section 75 (4) i.e. household, industrial and commercial waste or any such waste.

Section 33: - makes it an offence to carry out unauthorised or harmful deposition, treatment or disposal of waste. Penalties with regard to the offence are set out in the Act and for activities associated with non-special wastes these are:

- on summary conviction; imprisonment for a term not exceeding six months or a fine not exceeding £20,000.00, or both;
- on conviction or indictment: imprisonment for a term not exceeding two years or an unlimited fine, or both.

Section 34: - creates a statutory duty of care.

Sections 35 - 44: - introduce waste management licenses

Although the EPA was enacted in 1990 it did not have an immediate affect due to the fact that it is a statutory framework and it therefore requires secondary or subordinate legislation such as Regulations and statutory guidance to implement the detail. The Duty of Care came into force in 1992 but it took until 1994 for Regulations which dealt with the licensing aspect.

The Waste Management Licensing Regulation 1994.

During the consultation period for these Regulations, British Waterways was involved in protracted lobbying for Regulations which recognised the requirements of their operations. It was realised at an early stage in the process that lobbying by British Waterways alone was not going to achieve the desired result and it was therefore decided to pull together an industry group made up of navigation and drainage authorities, dredging contractors, consultants and Regulators. The Group was facilitated independently by CIRIA (Construction Industry Research and Information Association). This approach proved worthwhile as the outcome, although not perfect, did result in some favourable changes to the Regulations.

Following on from the lobbying, the Group, through CIRIA, produced a best practice guidance manual, "Guidance on the disposal of dredged material to land", CIRIA Report 157 which was published in 1996.

The 1994 Regulations resulted in significant changes to the way that dredging disposal operations were regulated. Waste disposal licences were converted to waste management licences and this was associated with changes to the licence conditions. The main changes being:

- The need for a license holder to be a "fit and proper person"
- The need for technically competent managers
- Payment of licence fees for annual subsistence, modifications, transfer & surrender
- The need to apply to surrender a license
- A schedule of activities which are exempt from licensing

The dredging industry has managed to operate within the requirements of the Regulations but it is not ideal having to comply with legislation which is intended for commercial landfill operations. At the end of the day, it results in facilities and operations being over regulated.

Post 1994

There are a number of EU Directives which are having, or will have, an affect on the disposal of dredgings. Examples of these Directives are:

- The Council Directive 1999/31/EC on the Landfill of Waste
- The EC Nitrates Directive 91/676/EEC

The Council Directive 1999/31/EC on the Landfill of Waste (The Landfill Directive)

In light of all the effort put into lobbying for the 1994 Regulations, it was somewhat of a disappointment when Defra published their first consultation paper on the implementation of the Landfill Directive in October 2000 (followed by a second consultation paper in August 2001). The proposals failed to offer a pragmatic approach to the regulation of the disposal of dredged material. Again it required British Waterways and other interested bodies to respond vigorously to all the relevant consultation documents. This time round, a formal group was not set up but there was regular communication between various parties to ensure that there was an element of consistency in the responses.

European Directives tend not to be explicit in the way that they are worded which allows member states to put some of their own interpretation into local regulations. This can have both advantages and disadvantages but, at the end of the day, it leads to inconsistent regulation throughout the EU.

Article 3.2 of The Landfill Directive contains two exclusions which are of particular relevance to dredging disposal activities: -

- *the spreading of sludges, including sewage sludges, and sludges resulting from dredging operations and similar matter on the soil for the purposes of fertilisation or improvement* (first indent of Article 3.2/ Regulation 4(a)); and

- *deposit of non-hazardous dredging sludges alongside small waterways from where they been dredged out and of non-hazardous sludges in surface water including the bed and its subsoil* (third indent of Article 3.2/ Regulation 4(c)).

The second of these exclusions consists of a very general statement which; both Defra and Scottish Executive have interpreted very specifically by linking the exclusion to the Paragraph 25 exemption of the Waste Management Licensing Regulations 1994.

Paragraph 25 exemption of the Waste Management Licensing Regulations 1994

25.—(1) Subject to sub-paragraphs (2) to (4) below, the deposit of waste arising from dredging inland waters, or from clearing plant matter from inland waters, if either—
 (a) the waste is deposited along the bank or towpath of the waters where the dredging or clearing takes place; or
 (b) the waste is deposited along the bank or towpath of any inland waters so as to result in benefit to agriculture or ecological improvement.

(2) The total amount of waste deposited along the bank or towpath under sub-paragraph (1) above on any day must not exceed 50 tonnes for each metre of the bank or towpath along which it is deposited.

(3) Sub-paragraph (1) above does not apply to waste deposited in a container or lagoon.

(4) Sub-paragraph (1)(a) above only applies to an establishment or undertaking where the waste deposited is the establishment or undertaking's own waste.

> **Paragraph 25 exemption of the Waste Management Licensing Regulations 1994 (cont.)**
> *(5) The treatment by screening or dewatering of such waste as is mentioned in sub-paragraph (1) above—*
> *(a) on the bank or towpath of the waters where either the dredging or clearing takes place or the waste is to be deposited, prior to its being deposited in reliance upon the exemption conferred by the foregoing provisions of this paragraph;*
> *(b) on the bank or towpath of the waters where the dredging or clearing takes place, or at a place where the waste is to be spread, prior to its being spread in reliance upon the exemption conferred by paragraph 7(1) or (2); or*
> *(c) in the case of waste from dredging, on the bank or towpath of the waters where the dredging takes place, or at a place where the waste is to be spread, prior to its being spread in reliance upon the exemption conferred by paragraph 9(1).*

It is the view of British Waterways, and the rest of the navigation industry, that the exclusion within the Directive would have been more explicit if it had been intended to be as prescriptive as the UK Government's regulations. The linking of the exclusion to the exemption has been perceived by the industry to be "gold plating" and not in keeping with the spirit of the Directive. The narrow interpretation of the exclusion within the UK regulations prevents its application to dredging disposal facilities which are alongside the waterways and currently regulated by Waste Management Licences.

Liquid Waste

One of the requirements of the Landfill Directive is the banning of liquid wastes at landfill, unfortunately neither the Directive or the UK Regulations define what constitutes a liquid waste. Defra put forward a working definition which has now been taken on board by the Regulators in England, Scotland and Wales. The definition states that liquid waste is any waste that has either of the following characteristics.

- *Any waste that near instantaneously flows into an indentation void in the surface of the waste.*
- *Any waste (load) containing free-draining liquid substance in excess of 250 litres or 10%, whichever represents the lesser amount.*

Although the principle of the ban is supported, in reality the tests have no scientific basis and they cause dredgings, which are generally acknowledged as being sludges, to be classified as liquid waste.

Where dredgings have been identified as a liquid waste they will have to be pre-treated to produce a physically stable material before they can be accepted by a landfill site. Although representation has been made to Defra, and the Regulators, about the suitability of the tests, they show no intention of considering alternatives.

Treatment

The treatment of dredgings is not a new concept, British Waterways have been involved in a number of trials over the years but their success has been limited due to the difficulties of producing small-scale plant which is easily mobilised for use on inland waterways. In addition, the cost of treatment has generally been high and sometimes excessively so. To date, the most effective treatments have been those which are simple and based on known technologies which have been developed and proved in other industries.

Historically, the main reason for treating dredgings has been to produce a material which is easier to landfill. Various types of treatment are available e.g.:

1. Physical stabilisation.
2. Volume reduction.
3. Chemical stabilisation.

The most common method of treatment has been physical stabilisation achieved by mixing the sediment with a product such as lime, cement or pfa.

As the legislation controlling the landfilling of wastes becomes more complex, specialist contractors are looking more seriously at the processing and treatment of dredgings. Currently the amount of plant available is limited and the complexity of it is variable as different contractors approach the problem from different angles. Plant development is at an early stage and over time it is expected that it will become more efficient and effective. There is still a premium cost associated with treatment and even taking into account volume reduction and the re-use of material, it may still be greater than disposal at landfill. Over time, it is expected that the cost differential will reduce, or disappear, as landfill costs rise and plant becomes more efficient.

The Nitrates Directive

Nitrate Vulnerable Zones (NVZs)

The implementation of the Nitrates Directive into UK law resulted in dredgings being defined as a Nitrogen fertiliser. When the NVZ Rules were revised in 2002, the area of England designated as an NVZ increased from 8% to 55%. Unfortunately for British Waterways, the bulk of their canal network is within the designated zone as can be seen in Figure 1.

Figure 1. Nitrate Vulnerable Zones in England

This definition of dredgings as a nitrogen fertiliser has had a serious impact on British Waterways, as it means that the application of dredgings to agricultural land must not exceed the annual crop requirement of fertiliser nitrogen. In determining the application rate, the assessment needs to take account of crop uptake, crop residues and organic manures. Operationally, this has resulted in the thickness of dredgings spread to land being reduced and, in some instances, the reductions have been significant. British Waterways have been in discussion with the Environment Agency and Defra to try and persuade them that dredgings should not be defined as a nitrogen fertiliser and for them to consider a more scientific approach for the spreading of dredgings to agricultural land. The current guidelines have been arrived at without an understanding of how the nitrogen in dredgings reacts when removed from a watercourse and spread on land. It has now been agreed that trials should take place to provide evidence of what happens when dredgings are spread to land. The trials are expected to start shortly, led by British Waterways, but with support and involvement from Defra and the EA. The intention is that the results will be available in time for them to be taken into consideration when the NVZ Rules are reviewed in 2005.

Lobbying

As discussed above, British Waterways and other organisations such as the Broads Authority have lobbied strongly for Regulations which are practical and relevant. Navigation interest groups such as the IWA (Inland Waterways Association) and AINA (Association of Inland Navigation Authorities) have also submitted strong representations. Following the consultation exercise in 2001, it was decided to request a meeting with Defra and the EA to raise the awareness of the problems and attempt to convince the regulators face to face. The request was eventually accepted and there has followed a series of meetings involving various departments of Defra, the EA, British Waterways and the Broads Authority. The early meetings failed to make any impact but following the provision of reports from BW, the Broads Authority and independent lawyers, as well as face to face meetings with Defra and the Agency lawyers and intervention from BW's Directors, progress began has been made.

Hydrodynamic Dredging

This is a generic term covering the re-distribution/relocation of sediments within the same, or hydraulically interconnected water bodies. The methods used in the process vary and encompass sediment agitation, pumping, and physical/mechanical transfer.

It is a technique which lends itself to riverine and estuarine environments rather than in canals where its uses are more limited. As the legislation regarding waste disposal become more complex it is a technique which is being considered more frequently. In order to ease the regulatory process associated with hydrodynamic dredging techniques, BW has entered into a Memorandum of Understanding (MOU) with the Environment Agency. The purpose of this MOU is to ensure effective cooperation between British Waterways (BW) and the Environment Agency (Agency) in dealing with dredging operations involving the re-distribution/disposal of dredgings within controlled waters (hydrodynamic dredging). The MOU (along with the underpinning guidance) provides a protocol for planning and undertaking dredging operations so that adverse environmental impact is minimised and the long term sustainability of the water system is not compromised and provides that the regulatory approach to the operations is clearly understood. There are two broad categories of hydrodynamic dredging: -

"In situ" techniques which involve agitation of sediment "in situ" so that it is carried downstream by the water flow, e.g.:
    a) water injection dredging
    b) harrowing, raking and ploughing
    c) use of grabs and backhoe to move sediment

"**Downstream**" techniques, used where there is insufficient water flow at the location to allow "in-situ" techniques, but where sediment can be transferred and released back to the same water body at a different location further downstream where water currents are high enough for dispersal, e.g.: -
    a) discharge from a split-hopper barge
    b) cutter-suction dredging where dredgings are sucked up and discharged downstream via a pipe or discharged from a split hopper barge.
    c) use of grabs and backhoe to move sediment

Current Concerns with regard to Regulation

As discussed above, the UK legislation controlling waste disposal at landfill is biased towards the regulation of commercial waste disposal facilities. As the legislation becomes more complex, and somewhat inflexible, those involved in dredging disposal activities are finding it more and more difficult to operate their facilities economically and are being forced to consider other disposal options which are environmentally unsustainable.

It has been noticeable over the years that policy makers and regulators in the UK generally have a limited understanding of dredging and dredging disposal operations. One of the misconceptions that the Regulators have is that they consider that dredgings are unsuitable for sustainable reuse e.g. for use in construction work.

At a recent meeting on sediments attended by practitioners and researchers into sediments; the workshop report stated that there were serious regulatory barriers to the effective management of sediments in general and dredgings in particular. Concerns were recorded with regard to inconsistencies and fragmentation in the regulatory process. It was noted that there was a lack of consistency between local and area offices of the Environment Agency regulation staff, which was exacerbated by high staff turn over, problems with fragmented assessment and individual interpretation varying from officer to officer (Apitz *et al*, 2002).

Potential Impacts

The primary impact on the navigation industry will be an increase in costs for dredging projects. This in turn will cause navigational bodies to restrict the level of dredging to an absolute minimum and increase use of hydrodynamic methods.

Elsewhere, avoidance of disposal of dredgings to landfills is likely to result in larger quantities of dredged material being deposited along the banks of the navigation channel. This in turn may impact on uses of riparian property.

In urban settings where there is limited scope for dredging disposal, the impacts may be far reaching. It is in urban settings that the most challenging dredged material is found, resulting from historic activities. The contaminants are often elevated and there is usually a greater occurrence of refuse such as shopping trolleys and traffic cones.

Further impacts on dredging activity may be caused by the Water Framework Directive; if it can be shown that dredging changes the ecological status class of the water body. For heavily modified waters there is a derogation for physical alterations made for which the artificial characteristics of the body serve "as long as all mitigation measures are put in place to the best approximation to ecological continuum". This means the Directive recognises the need to dredge for navigation purposes but best practise must be applied. This does mean that further lobbying, research and debate may be required to show that current practice is best practice for the maintenance of the ecological status of the water body.

Conclusion

The last 20 years has seen a significant change in the way that the disposal of dredgings has been regulated in the UK; from no control prior to 1988, to the current situation where complex legislation is in place. The implementation of this legislation has caused significant difficulties for those involved in the disposal of dredgings from inland waterways, mainly because it is biased towards the operation of commercial landfill. The problems have been exacerbated by the implementation of EU Directives into UK law, where government departments have made assumptions based on a lack of understanding and technical knowledge.

Drainage and navigation authorities must not be complacent; they must remain vigilant for signs of new legislation on the horizon and changes to that already in existence. Where changes are proposed, they must get fully involved in consultation exercises and lobbying for legislation which is relevant and practical. To help in this process, there is a need for these bodies to work more closely together in order to share best practice, to raise awareness of issues and to increase the power of lobbying.

References

Apitz, S., Burt, N., Cook, S., Fletcher, C., Lavender, J., Murray, L., Parker, A., Power, E.A., Wilkinson, H., White, S., Wright, T 2003' Holistic Sediment Management: UK goals, needs and approaches - workshop report,
In: Proceedings of Sediment Management: The Challenges of 2003, 9th & 10th April 2003. (http://www.soils.rdg.ac.uk/events_and_news/Workshop%20Report.pdf)

Construction Industry Research and Information Association, 1996. Guidance on the disposal of dredged material to land. Report 157, Construction Industry Research and Information Association.

Department for Environment, Food & Rural Affairs, 2002. Guidelines for farmers in NVZs – England, Department for Environment, Food & Rural Affairs.

Department for Environment, Food & Rural Affairs, 2001. Implementation of Council Directive 1999/31/EC on the landfill of waste. Second consultation paper, Department for Environment, Food & Rural Affairs.

IWA Head Office Bulletin - February 2001 - Issue No 49

Improve your chance of approval – beneficial use of dredged materials.

John Brien Harwich Haven Authority, Harwich, UK.

Introduction

The focus of this paper is the requirement under the Food and Environment Protection Act 1985 that the Department for Environment, Food and Rural Affairs (DEFRA) or the other appropriate regulators, consider practical, alternative disposal options when applying for consent to dispose of materials at sea. The aim is not primarily beneficial disposal but that of ensuring that the minimum amount of material is eventually disposed of at sea.

This approach is in line with that adopted in a wide range of disposal operations, often referred to as the *3 Rs*: *Reduce, Re-use, Recycle* – before considering treatment if necessary and disposal of what remains. This paper will briefly review the legislative framework and the present policies of Government departments, and will consider alternatives to disposal at sea under the broad headings of reduce, re-use and recycle. It will cover areas which are present DEFRA policy and also suggest some ways in which the approach could possibly be widened in the future.

Legislative framework.

The Food and Environment Protection Act 1985 (FEPA) provides the statutory means for the UK to meet its obligations under the London and OSPAR conventions, which aim to prevent marine pollution from dumping at sea. DEFRA administer the Act in England, with Scottish and Welsh Assemblies taking responsibilities within their jurisdictions. The Objective of Licence control is introduced in Clause 8.1;

In determining whether to issue a license a licensing authority,
    (a) shall have regard to the need-
        (i) to protect the marine environment, the living resources which it supports and human health; and
        (ii) to prevent interference with legitimate uses of the sea; and
    (b) may have regard to such other matters as the authority considers relevant.

Clause 8:2 requires that the Authority,
    ...in determining whether to issue a licence, shall have regard to the practical availability of any alternative means of dealing with (the materials for disposal).

It is DEFRA's stated policy that no waste should be disposed of at sea if there is a safe and practicable land-based alternative[1]. Since the end of 1998 most forms of disposal at sea have been prohibited, including toxic wastes, incineration at sea and dumping of sewage sludges. The only materials still being disposed of at sea are dredgings from ports and navigation channels, hence the need for these to be strictly controlled, and reduced to the minimum practical requirements.

As well as a policy of disposal minimisation, DEFRA seek to use as much clean dredged material as possible in beneficial disposal options and refers to beach nourishment and saltmarsh or mudflat enhancement. The MCEU web site also contains a flow chart showing how materials can be selected and used in beneficial schemes[2] and makes provision for a separate licence category with reduced application costs for such applications.

The discussion regarding what constitutes capital and maintenance dredging operations has continued for a number of years. The difference, though in theory down to whether the area has been dredged within a specified time, does tend to be reflected in different material types.

The official definitions used by DEFRA at present are:

> **Capital dredgings:** Material arising from the excavation of the seabed, generally for construction or navigational purposes, in an area or down to a level (relative to Ordnance Datum) not previously dredged during the preceding 10 years.
>
> **Maintenance dredgings:** Material (generally of an unconsolidated nature) arising from an area where the level of the seabed to be achieved by the dredging proposed is not lower (relative to Ordnance Datum), than it has been at any time during the preceding 10 years, *or* from an area for which there is evidence that dredging has previously been undertaken to that level (or lower) during that period.

In practice this means that maintenance dredgings will be materials which accumulate reasonably quickly in areas which require maintaining at a set depth. They are therefore, much more likely to be fine grained; silts, clay and fine sands, rather than coarser materials. Where the coarser materials; medium to coarse sands and gravels, are removed in maintenance dredging operations, there is generally a ready supply of projects eager to receive them, from beach nourishment to reclamation and supply of aggregate for concrete production. It is the finer grained materials, particularly the clays and silts, which are produced in large volumes and present the greatest problems in finding suitable alternatives to sea disposal.

Alternatives to disposal at sea.

"Reduce"

This approach, the first of the *3R's* is the one which at present, DEFRA has no mechanism for considering but if they are to seek *all* the alternatives to disposing of material at sea, then

---

[1] Taken from the Marine Consents Environmental Unit (MCEU) web site, www.mceu.gov.uk/MCEU_LOCAL/FEPA/MENU-IE.HTM
The MCEU is a joint unit from DEFRA and the Department for Transport (Ports Division) which handles disposal licence and Coast Protection Act consent applications
[2] www.mceu.gov.uk/MCEU_LOCAL/FEPA/FEPA-beneficial.htm

perhaps this should be recognised as a real option. There are a number of questions which could be asked:

What would happen if we stopped or reduced the amount we dredge?

It is possible to be trapped in the assumptions that because maintenance dredging was needed at one stage, it always will be. Sediment sources can change, works in other parts of an estuary or system can effect the accretion patterns. Increases in deep draughted vessel passages may be keeping more material in suspension and fresh water inputs may have altered the balance in the harbour. There can be a variety of reasons for maintenance dredging requirements to change and there have been industry examples of ports which found they could dramatically reduce their maintenance dredging commitment with little impact on accretion levels or navigational access.

Other means of reducing the volumes removed could include reviewing the need for full under keel clearances (UKC) within a sheltered harbour or channel, or in areas where vessels would be manoeuvring at slow speeds. Areas which are regularly dredged are likely to have a disturbed and less dense layer on the bed, which may or may not be properly represented by echo-sounding. There are ports which operate with a "navigable mud" layer of low density, which is not considered to be restricting to deep draughted vessels. Such consideration could contribute to reductions in the amount dredged and disposed, particularly if adopting a reduced UKC over soft materials led to an increase in ship generated erosion of the bed.

Putting forward a reduction in dredging volume is unlikely to be seen as a realistic alternative in itself, as if such a reduction were practically possible the applicant would be using the lower figures. However, a reference to a study showing that the minimum possible volumes were being applied for and that all means of reduction had been considered, should be seen as producing the desired result – the minimum disposal at sea.

Can we obtain the required results without dredging and therefore needing to dispose of the material?

Improvements to navigable depth which can be achieved using methods not involving dredge and disposal are beneficial in a number of ways. With no requirement to dispose of material, there are no concerns regarding effects at the deposit site. In many situations the material to be dredged was originally deposited as part of a dynamic estuary and near-shore system of sediment transport. The removal of material to a remote deposit site has the potential to upset the balance of erosion and accretion in the area, so retaining it in the locality would be an advantage.

The commonly used methods which provide these benefits are ploughing or bed-levelling, and the use of Hydrodynamic dredging methods such as Water Injection Dredging (WID). In both these methods, material is moved from areas where it is causing shoaling either into existing deeper water, or in a manner which leads to it becoming entrained in the normal tidal currents and dispersed. It is clear that deliberate maximisation of ploughing or WID will inevitably lead to a minimisation of material to be disposed of at sea. It could be argued that DEFRA should consider that if maximum use was being made of these techniques then the requirement to seek alternative disposal options and minimise sea disposal was at least in part, being met.

"Re-use"

The heading of re-use, covers the bulk of the alternative disposal options currently undertaken or considered. To re-use the material implies use in its as-raised state without significant modification and therefore needs to be considered separately for the different materials available.

- **Coarse Sand and Gravel** - In the unusual cases where maintenance dredging involves coarse sand or sand and gravel mixtures, there will be such a range of potential alternatives that DEFRA is unlikely to licence disposal to sea except in exceptional circumstances. Material is in demand around the coast for beach nourishment and sea defence projects, many of which will be able to adapt their works programme or even design criteria to accept the material which will be raised. Reclamation schemes, which often use associated capital dredging works, or the import of suitable materials from sea or land sources, could also use material from maintenance dredging projects. The long lead times required for grant aided coastal works or for reclamation would mean that such projects would be more likely to benefit from maintenance dredging regimes which produce a reasonably predictable material both in volume and specification. These coarse materials can also find a place in Habitat creation or enhancement works, normally providing protection for vulnerable sites or areas for nesting birds or shingle flora.

- **Medium and Fine Sands** - Finer grain materials are less suitable for use in structural reclamation, though if it is well contained there are uses as bulk fill or reclamation remote from quay structures. Although not employed so much in the UK, marine sourced sands are used extensively in Europe in a wide range of construction projects. Sand is used in the UK for beach and dune nourishment, particularly in more sheltered locations or in combination with groynes or breakwaters.

    As with the coarser materials, sand can be involved in recreating or improving habitats, often together with fine sediments. There are examples of sand or gravel banks retaining and protecting existing habitats, or providing short term restraint behind which new fine materials can consolidate and habitats develop[3].

- **Silts and Clays** - A large proportion of maintenance dredgings are either solely silt and clay muds or a mixture of silt and fine sands. These are the materials for which there is most pressure to find alternative disposal options and also the ones which are most difficult to achieve. Extra pressure comes from the fact that these are the materials that may carry any contaminants present, bound to the fine particles. In some European ports the sand and silt fractions are separated, allowing re-use of the clean sands and contained, on-shore disposal of the silts.

    Alternative disposal by re-use of silts and clays has generally been in habitat creation and enhancement projects, flood prevention or sediment replacement schemes, and disposal to on-shore facilities.

    Habitat enhancement schemes have become more common as the requirements of the Habitat Regulations have led to mitigation and compensation works for capital projects. Although the application of the Regulations to maintenance dredging is a matter of

---

[3] Such as theTrinity III Extension foreshore enhancement works on the Orwell: 2003.

current debate, the existence of suitable trial schemes and projects has led to habitat enhancement being seen as a real alternative disposal option. At the two ends of the scale; dredgings from small marinas have been pumped directly onto saltmarsh and mudflats[4], and the proposed managed realignment site on Wallasea Island on the River Crouch in Essex is planned to take 500,000m³ of maintenance dredgings[5]. The aim of these schemes is to improve existing, or to produce new sustainable and productive habitats. They are usually subject to intensive survey and monitoring to establish their habitat value and to ensure they are providing the benefits predicted.

Flood protection can be either a primary benefit sought from schemes, or more often, it is a secondary outcome. Fine sediments are unlikely to provide much in the way of structural reinforcement to a flood defence, but in situations where foreshores have been raised, the severity of wave attack will have been reduced. There are locations where defences which will erode away in the medium term have been preferred, as this would not restrict future options, once shoreline management plans or CHaMPs were finalised.

Sediment Replacement schemes have been put in place in a number of areas, often where the maintenance dredging may be contributing to on-going erosion of mudflats or saltmarsh areas. Analysis and modelling which shows that the mechanism exist for this increase in erosion, can also demonstrate that material can be fed back into the system – sometimes referred to as 'trickle-charging'. Sediment can be placed directly onto the foreshore by pumping or shallow dumping or more usually, placed into the water column to be distributed by the normal hydraulic processes. Continuing monitoring and review is normally needed to ensure that the volumes reintroduced are adequate but not excessive, allowing for the inevitable inefficiency of a river or estuarine system in feeding material to specific areas. Details of the Harwich sediment replacement scheme can be found in Appendix 1.

On-shore disposal facilities have been developed generally for simple logistical reasons, such as the long distance to a suitable licensed sea disposal site. This is obviously the case with inland waterways, but also applies to areas within large estuaries, such as the Thames. The large areas needed for onshore disposal tend to limit their use to areas with reasonably modest maintenance dredging needs, unless specially designed sites, like the Slufter in Rotterdam can be constructed[6].

On-shore disposal has been considered in the past simply as a potentially expensive alternative to dumping at sea. However, a number of sites have been designated by the conservation authorities for their value for birds and other wildlife. As sites of Special Scientific Interest (SSSI's) they are managed to maintain this conservation value, with control on water levels and areas under use at any time. As such, the regular placement of dredge materials in these sites is playing an essential part in their conservation management. These sites could legitimately be viewed as beneficial disposal options in their own right. Details of the Westminster Dredging disposal site at Cliffe on the River Thames can be found in Appendix 2.

---

[4] http://www.cefas.co.uk/decode/use.htm
[5] http://www.defra.gov.uk/wildlife-countryside/ewd/wallasea/Leaflet.pdf
[6] A 260 ha. specially constructed site, originally for the disposal of medium level polluted sediments, designed to contain approximately 100 Mm3. The need for higher level polluted sediment disposal has reduced, so all Rotterdam's non-sea disposal is now carried out here. Overall disposal volumes have also reduced, giving the site an increased effective life.

Fine sediments can have a role in capping or sealing sites for the disposal of polluted sediments or land sourced waste materials. Sediments with a high percentage of clay will provide a layer of restricted permeability and reduce leaching of pollutants into the ground water.

"Recycle"

As in other disposal minimisation programmes, recycling covers operations which take the material and change it in some way to make it usable as it is, or as a raw material in a different process. Recycling normally requires some energy input and should therefore be considered only after the possibilities of reduction and re-use have been exhausted. The principal alternatives to sea disposal by recycling are the use of sand and gravel in concrete production and processing silts into lightweight aggregate or blocks. Gravel is an unusual material to be raised in maintenance dredges and would generally be used for reclamation or shore protection schemes. Sand however is dredged and sold for concrete or mortar production in various ports in the UK and Europe. In some places it is raised in a mixture with silts and these need to be separated both to provide clean construction sand and to remove contaminants for further treatment. A number of trials have been carried out to find uses for the remaining silt materials, specially where they carry contaminants. The fixing of these pollutants in stable materials can be achieved by processing the silts and heating in furnaces to produce lightweight aggregates for use in specialist concrete, blocks or roof screeds. The energy requirements for these processes can be high, making the products expensive and the market tends to be reluctant to use materials made from possibly polluted 'muds', however good the environmental credentials may be.

Assessing alternatives and beneficial uses – who decides?

When an applicant completes question 15 of the MCU3 form for a disposal licence from DEFRA, they are required to;

.. state the results of your investigations to identify alternative means of disposal (including beneficial uses) or treatment of the material, together with costs of all the options, including deposit at sea.
Failure to complete fully this section will result in the form being returned.

This statement raises a number of questions;
- How far do you need to go to consider alternatives?
- Can they be ruled out simply on financial terms?
- If so, how much extra cost is unreasonable?
- In terms of beneficial use, who do they need to benefit?

There is at present a startling lack of guidance on these matters. It has been suggested to applicants in the past that they should be allocating for capital projects up to 10% of the predicted cost of the works to fund extra beneficial disposal options, but this does not appear to be a consistently applied approach. Given the long time periods required for planning approvals or environmental investigations, it would seem likely that many alternative disposal options will need to be promoted by others – local authorities, flood prevention agencies or conservation bodies, working in conjunction with the harbour authorities or others planning to carry out dredging. This does require good co-operation and the possibility of financial planning and allocation of funds over considerably longer than the usual 1 or 2 year budget cycles. DEFRA is attempting to provide a data base of potential disposal sites and sources of dredge material, but this is at an early stage and will require much greater input and involvement from all the participants if it is going to operate successfully.

One of the difficulties in collaborative working for these projects is that the conservation bodies and local authorities in particular are facing severe financial constraints and are reluctant to commit themselves to work they consider DEFRA may require a commercial port operator to fund. Comparisons with Europe are also complicated by the fact that most of the port operators and harbour authorities outside the UK receive funding from central and regional government, which enables them to carry out these projects without imposing a significant burden on the port or its users. In the Port of Hamburg, maintenance dredging material is treated to separate the less polluted fine sand fraction before the remaining silts are treated and de-watered to be disposed of at an on-shore facility. The running costs of this operation, including the dredging, are approximately €30/m3 (£20/m3) and about €60M (£40M) was invested to set up the plant and disposal site. This will provide Hamburg with disposal options for many years but represents a level of expenditure UK ports would find hard to match.

The question of which party a beneficial scheme may be thought to benefit can further add to the difficulties in assessing these options. A low cost disposal in *any* location could be seen to be beneficial to those carrying out the dredging, but 'beneficial disposal' as such implies a wider community benefit, and generally carries an unspoken assumption in favour of conservation or flood defence improvements. Even when a conservation interest is being promoted the legislation which considers designated sites and species above others can lead to tensions. Sediment replacement programmes may be maintaining or even restoring designated inter-tidal mud flats for feeding birds and English Nature and RSPB may support erring on the side of over provision if the science is uncertain. In the same area there may be fishing interests, commercial and recreational, who would be concerned for smothering of sub-tidal benthos and fishery resources, and thus take a very different approach.

In many cases the way forward will be by agreeing to monitor and report on the development of the sites or the programme, including a commitment to review and amend the schemes if necessary. Such agreements have been a feature of recent capital dredge and development schemes, but are likely to become more common in relation to maintenance dredging operations, particularly where these are occurring within or adjacent to SPA's or other designated sites.

The title of this paper may be rather misleading as considering alternative and beneficial disposal options will not simply *"Improve your chances of approval.."* – such a consideration is, and will increasingly become absolutely essential. There is a wide range of possible alternatives to sea disposal, some of which the licensing authorities may not be currently considering. It should be remembered that it is reducing sea disposal that is the main thrust of the legislation and that alternative options should be sought before focusing on how beneficial they may be. The way forward must involve continuing engagement and debate about these issues at the earliest stages and a readiness to consider new approaches which may arise.

Appendix 1.

Sediment Replacement Programme; Harwich; UK.

The sediment replacement programme at Harwich was introduced following the 1998-2000 capital dredging project. The studies demonstrated that a larger proportion of the sediment which comes into the harbour from the sea would settle in the berths and approaches and have to be dredged. Prior to the deepening, this material would have passed through the harbour and entered the rivers, feeding into the dynamic interchange between water column, sub-tidal bed and inter-tidal areas. Dredging this material and disposing of it outside the harbour would therefore reduce the volume of material entering the rivers and could lead to increases in erosion.

The sediment replacement programme takes material from the lower harbour and places it back into the system by pumping it overboard at 6 locations in the rivers, and by placing it down the suction pipe onto the harbour bed in areas which naturally scour. A total of approximately 660,000 wet tonnes of the clay-silt mixture is placed each year, which is about 10 times the amount that the studies predicted could be lost to the rivers. This over provision is because of the acknowledged inefficiency in this method of re-introducing sediment; if it could be reliably placed directly onto eroding inter-tidal areas a considerably reduced amount could be used. The management and monitoring of this programme is reported annually to a Regulators Group, set up by agreement and required by the dredge consent. This group is required to review and amend the programme in the light of the monitoring which presently includes bathymetric surveys, sediment type and benthic sampling, bird counts, fish surveys, suspended sediment monitoring and targeted detailed surveys adjacent to the disposal sites. Since 2000, the amounts have been increased and changed from sub-tidal to water column placement in the rivers, placement times have been altered in co-operation with local fishing interests and one site, at Holbrook Bay has been suspended following the discovery of Native Oysters in the vicinity. Future changes may include smaller more targeted placements, introducing additional sites further up-river or the use of water injection dredging.

Annual deposits in sediment replacement sites (wet tonnes)

Area	Erwarton East (A)	Copperas East (B)	Copperas West (C)	Holbrook Bay (D)	Orwell East (E)	Orwell West (F)	Harbour Shelf (G)
Nov. to Dec.	34,000	34,000	13,000	13,000	13,000	13,000	100,000
Jan to Feb.	35,000	35,000	13,000	13,000	13,000	13,000	100,000
Mar to Apr.	35,000	35,000	13,000	13,000	13,000	13,000	100,000
Totals	104,000	104,000	39,000	39,000	39,000	39,000	300,000

This programme was initially a requirement of the capital dredge consent but is also an alternative and beneficial disposal option for maintenance dredgings. It illustrates the way that even if it were to be decided that maintenance dredging operations were not covered by the Habitats Regulations, they can be caught up following a capital dredge, where the impact of the works may partly only be revealed by the subsequent maintenance.

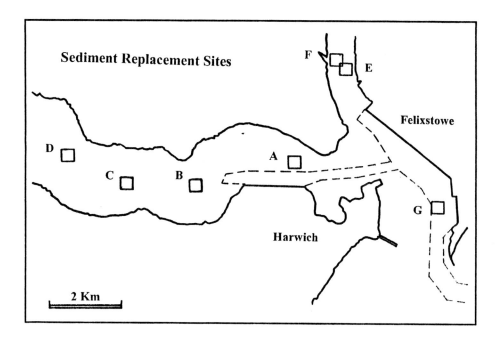

Appendix 2.

Cliffe Disposal Site; River Thames; UK

Cliffe is situated on the south bank of the Thames, downstream from Gravesend. The site was originally excavated as clay pits for cement manufacture which ceased in 1970. The whole site is approximately 230 ha, comprising 110 ha of saline lagoons, 27 ha of shallow freshwater pools and rough grassland and scrub. Saline lagoons are a rare habitat type in the UK and this site contains 10% of the UK total.

In the 1960's and 70's permissions were given to in-fill the pits by disposing of river dredgings, intending to eventually return the area to agricultural use. More recently, the conservation value of the present site was recognised and the possibility of partnership management explored. Those involved included the local authorities and interest groups, and in particular, the RSPB and Westminster Dredging Ltd., the licensed site operator. Plans have been agreed which maintain a disposal site for up to 2.6M m3 of river dredgings and manage the operations over the next 40 years by controlling water levels, placement rates and habitat creation to support the range of birds and other wildlife which uses the site. The continued disposal of material and the associated volumes of saline water are an essential part of maintaining the site. Volumes disposed of have varied between 75,000 and 200,000 m3 annually in recent years.

The site is being developed as an integral part of the wider Thames Gateway wild heritage area and will contain educational facilities and a visitors centre. RSPB intend it to be a *"Flagship nature reserve and the focus for visitors to the RSPB's North West Kent reserves"*. Recent visitor reports have recorded up to 63 bird species seen on one day including little egrets, marsh sandpipers, spoonbills, and avocets.

There are a number of other sites on the Thames which together take the total volume of on-shore disposal which is required. The position of the Port of London Authority (PLA) in approving dredging on the Thames is not typical of the UK but this site does illustrate how on-shore disposal can be beneficial both in terms of economics and conservation interests.

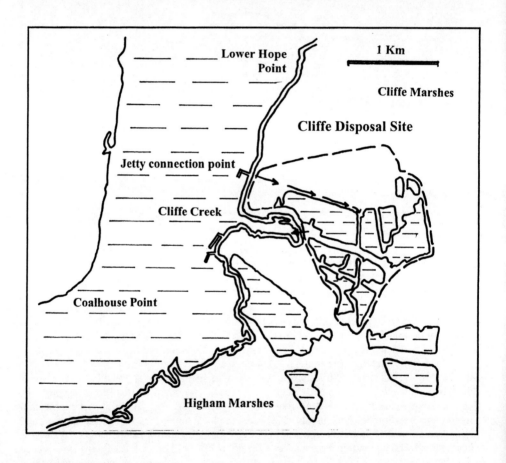

Hydrodynamic and Non-Dredging Solutions

Paul Hesk, Business Development Manager, Van Oord UK, Newbury, UK

Introduction

Whereas the majority of Ports and Harbours around the world require some form of maintenance dredging, various factors can contribute to often make it an inefficient and therefore relatively costly operation. The dredging of silts by conventional means, such as with a trailing suction hopper dredger (TSHD), is often inefficient due to the majority of the volume removed being water rather than solid. When one considers that a high proportion of the costs of dredging with such TSHDs arise from the transportation of the material, Ports and Harbour Authorities are paying a high price for the movement of the large quantities of water contained within these sediments.

Hydrodynamic Dredging methods have been developed to provide either a stand-alone solution for the maintenance dredging of silts and fine sand sediments or to complement conventional dredging techniques. The methods of hydrodynamic dredging rely on the deliberate re-suspension of fine sediments from the river or sea bed such that the material is removed from the dredging area using natural transportation processes. The principle transportation medium for this material is the water column itself.

The three main processes of Hydrodynamic Dredging are as follows:

- The injection of large volumes of water, at low pressure, into the settled sediments to create a density current (Water Injection Dredging).
- The stirring of the settled sediments, generally by mechanical methods (Agitation Dredging).
- The use of Conventional Dredging equipment to release sediment into the upper layers of the water body. (e.g. overflowing).

The disposal of dredged material at sea within the UK is regulated by the Food and Environment Protection Act (FEPA) 1985, implemented by the Department of the Environment, Food and Rural Affairs (DEFRA). (The Dumping at Sea Act 1996, implemented by the Department of Communications, Marine and Natural Resources, provides the equivalent legislation in Ireland).

This legislation has jurisdiction over operations which require both the removal of sediment from the seabed and subsequent re-deposition from a vessel. As both the first two of the three methods above do not require such re-deposition from a vessel, they fall out with the coverage of the current legislation and, as a consequence, no FEPA (or equivalent) disposal licence is required.

Whereas hydrodynamic processes are relatively cheap and easy to use, a potential problem has always been that, once mobilised, the control of the sediment is limited and its final destination difficult to predict.

The purpose of this paper is to provide an update on the developments being made in the prediction of the production and transportation distances achieved through the first of the processes described above, i.e. Water Injection Dredging.

Water Injection Dredging – An Overview

The Water Injection Process

The process of Water Injection Dredging was developed in the 1980's; the patent for which, through a series of mergers and acquisitions, is now operated by Van Oord NV. Van Oord owns several Water Injection Dredgers which operate throughout the world, but particularly in Western Europe, India and South America.

The Water Injection Dredging process is generally applied on soft seabed deposits, for instance silts and fine sands, where they will be re-suspended and subsequently re-absorbed within the existing sediment transport system.

Water Injection Dredging puts the seabed layer into suspension by injecting water into the subsoil, thus fluidising it (See Figure 1). This fluidised top layer, the thickness of which (typically between 1m and 3m) is highly sensitive to both the soil parameters and the distance between the injector pipe and the seabed ("stand-off distance"), then behaves as a density current. The low injector pressure (approx. 1 bar) and keeping a small stand-off distance are key factors in minimising turbidity. The optimum stand-off distance, which depends upon the settled sediments' particle size distribution and permeability, is often determined through practical experience.

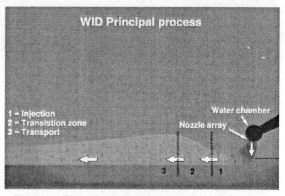

Figure 1. The Water Injection Dredging Principle

The properties of the induced fluid mud layer depend on a large number of factors, but of prime importance are the particle size and viscosity of the sediment.

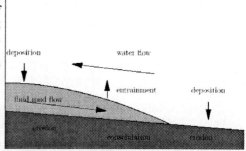

Figure 2. The Material Exchange Processes

The density current thus formed flows under the combined effects of gravitational, hydrostatic, current and friction forces.

The fluid mud layer flows under the influence of gravity caused by the gradient in either the rigid bed or the fluidised layer / water interface. The fluid mud layer may gain material from the rigid bed through erosion or, conversely, material may be transferred to the rigid bed through consolidation.

Similarly, the fluid mud layer may also gain or lose material by deposition from or entrainment to the overlying suspension layer respectively. In general, for a fluid mud layer generated by Water Injection Dredging, the processes of entrainment (loss of material to the water column above) and consolidation (loss of material to the rigid bed below) are the most important.

In addition to the gravitationally derived forces detailed above, the flow of water also imposes forces on the fluid mud flow. The force generated by the flow of water, albeit offset slightly by bed friction, is generally the main one acting on the fluid mud.

Environmental Impacts
Van Oord, in co-operation with various Clients and Research Institutions (including Delft Hydraulics and HR Wallingford), have undertaken a number of research projects throughout the world to demonstrate the impacts of Water Injection Dredging on the Environment. During these Projects extensive on site monitoring systems were installed to measure numerous properties, such as soil characteristics, turbidity, dispersion and current velocities, before, during and after dredging operations. In all of these research projects the water injection dredging process resulted in well-defined density currents and interfaces with the surrounding water.

As more experience has been gained in the use of Water Injection Dredging, it has become possible to regulate its use to such an extent that layers of sediments can be removed without unacceptable, adverse environmental impacts. For instance, the majority of water injection dredging within the Thames Estuary takes place only on Ebb tides to prevent the possibilities of deposition of material within environmentally sensitive areas.

The environmental advantages of Water Injection Dredging, assuming the sediment has the required characteristics, can be summarised thus:

- The dredged material is retained within the river system. During transportation the density layer is generally moving towards deeper waters at nature's own pace and therefore has little effect on the surrounding water.
- Only the bottom of the water column (typically between 1m and 3m as described above) is affected during dredging.
- The deposition of the dredged material takes place gradually, preventing any shock effect on the sea or river bed.
- Lower energy consumption per m^3 of material dredged.

Prediction of Production Rates and Transport Distances achieved by Water Injection Dredging

Introduction
As detailed above, a potential problem with all types of Hydrodynamic dredging, and in particular Water Injection Dredging, was that once mobilised the control of sediment is limited and its final destination difficult to predict.

Historically, any investigations into the effects of Water Injection Dredging would have necessitated detailed and expensive three-dimensional modelling of the water motion and sediment transport. To obviate this need Van Oord, in co-operation with Delft Hydraulics, has through extensive research developed a relatively quick and simple model, the "Rapid Assessment Tool" ("RAT") to assess these effects.

The Rapid Assessment Tool ("The RAT")
The Rapid Assessment Tool is a two layered model, depicting both the overlying water and fluid mud layers (See Figure 3). The overlying water layer is modelled using an analytical solution of the tidal averaged 2DV continuity and momentum equations. The fluid mud layer is represented using a one-dimensional, depth integrated numerical model. Using this model, production rates and transportation distances for Water Injection Dredging applications can be easily assessed using readily available data.

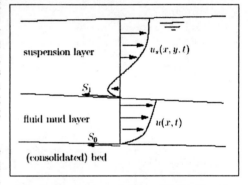

Figure 3. Layer modelling with the Rapid Assessment Tool

The Rapid Assessment Tool is particularly suited to the studies of production rates and transportation distances in estuarine waters. In estuaries, the direction of the water flow depends on both tidal action and fresh water discharge. In addition, the interaction of fresh and salt water and tidal currents is prone to causing fluctuating velocity profiles. The Rapid Assessment Tool is capable of calculating these variable profiles.

The input of the Rapid Assessment Tool consists of a number of variables which can be broadly divided into the following three categories:

- Properties of the estuary (depth, slope, tidal period, fresh water discharge, width etc.)
- Properties of the fluid mud flow (density, yield stress, friction coefficients etc.)
- Numerical parameters (parameters used to perform the actual calculation e.g. simulation time, time step etc.).

The output consists of two graphs, the fluid mud layer thickness at several moments in time and the production rate as a function of the distance from the location of dredging operations.

The upper graph predicts the behaviour of the density layer both as functions of distance along the channel from the dredging operation and time. Figure 4 shows that the density layer increases in thickness the further it gets from the dredging site (mainly from erosion and deposition from the underlying and overlying layers respectively) but reduces, as would be expected, over time.

The lower graph predicts the quantity of fluid mud passing at a distance beyond the dredging site. It can be seen from the lower graph in Figure 4 that the production rate is stated in units of m^2. This is due to the RAT model being only two-dimensional. In reality, the width of the density current will be approximately the same width as that of the injection beam on the dredger (12m for the water injection dredging vessel "Jetsed"). Therefore a cubic metre production rate for the fluid mud layer can easily be calculated from the lower graph in Figure 4.

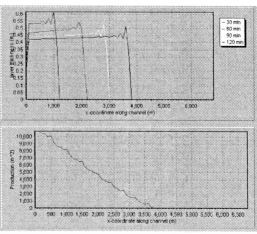

Figure 4: Example Output from the Rapid Assessment Tool

The RAT model's numerical output also allows graphical representation of the following parameters:
- fluid mud quantity against time
- Entrainment (loss of material to the water column above) and Consolidation (loss of material to the rigid bed below) quantities against time.

This numerical output is also suitable to be used in the production of visual reproductions of the fluid mud flow should the application so require it.

Case Study
In order to both assess and demonstrate the output from the Rapid Assessment Tool, Van Oord have carried out a case study analysis of the Water Injection Dredging works regularly undertaken at Custom's House Quay on the River Thames.
The input parameters for the model were supplied by the Port of London Authority, Dredging Research Ltd and Van Oord.

Input Parameters
The following main categories of input parameters were determined:

- A geometric and numeric description of the Fluid Mud Transport Channel
- The physical properties of the water, mud, solid bed and solids.
- A geometric and numeric description of the fluid mud layer.

The following key points regarding the input parameters and subsequent modelling should be noted:

- The Water Injection Dredging at Custom's House Quay takes place, in effect, on the slope of the main Thames Channel. As this slope is relatively steep, it is assumed that no material is lost through consolidation in this area. The RAT model is therefore used to assess the transport of the dredged material within the main channel only. Due to the irregularity of the main channel near the dredging area at Custom's House Quay, for instance there is a local depression nearby, a flat bed is assumed for modelling purposes.
- It is further assumed that there is no initial fluid mud layer present within the main channel prior to the Water Injection Dredging operations commencing.
- The assumed flow velocity and mixture density of a WID induced density current are shown in Figure 5. These values were measured in a trail undertaken in fresh water. When these values are modified to allow for the dredging at Custom's House Quay being undertaken in saline water, it is reasonable to assume a produced fluid mud density of $1100 kg/m^3$.
- One of the "RAT" model input parameters is the width of the estuary mouth. In this particular application the exact location of the estuary mouth is not critical to the model. A significant quantity of hydraulic data is available for the Thames near Shellhaven and therefore a cross-section of the river at this point is assumed as the theoretical estuary mouth.
- No information regarding the Bingham yield stress of the dredged material was available, therefore a value of 2Pa was assumed. (This is based on research within Dutch Harbours that has shown average values of this order for water – mud mixtures with densities of approximately $1100 kg/m^3$).
- Dredging was assumed to have commenced at the time of maximum ebb velocity and to have been of 1 hour duration.
- As the tidal water movement of the Thames is more significant than fresh water discharge, a well mixed (i.e. a small, vertical, salinity gradient) estuary and the absence of a salt wedge is assumed.

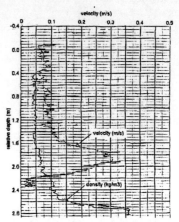

Figure 5: Fluid Mud Discharge Characteristics

RAT Output

The output from the "RAT" for this Case Study is shown in Figure 6. The model predicts that the fluid mud layer thickness, whist initially approximately 2m thick, is generally in the range of 1 to 1.5m thick. Although the fluid mud layer becomes thinner over time, the process is very slow. This reduction in thickness is due to a combination of losses caused by entrainment and consolidation. The upper graph also shows that the propagation of the fluid mud along the main channel slows considerably (to approximately 20 m/hr) after 120 minutes, culminating in a maximum transport distance of 1200m.

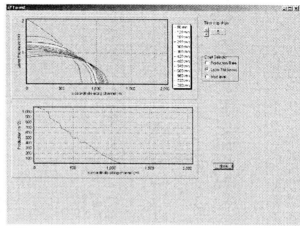

Figure 6: "RAT" output for Water Injection Dredging at Custom's House Quay, River Thames.

The model output predicts a two-dimensional liquid mud flow of 500m^2 at a distance of 500m from the source of the dredging. As previously described, this flow rate can be converted to a three-dimensional volume by multiplying the two-dimensional output by the width of the injection arm (12m). Therefore the model predicts a fluid mud quantity of 6,000m^3 (or a production rate of 6,000m^3/hr as the dredging duration is assumed as 1 hour) at 500m from the dredging source. By comparison of the in-situ and fluid mud densities, this converts to an in-situ volume flow rate of approximately 1,200 m^3/hr.

Analysis of the numerical output from the model allows the production of a graph representing the quantity of liquid mud over time (Figure 7). This graph shows the production of the fluid mud through the water injection dredging process for the first 60 minutes, followed by the gradual reduction due to entrainment and consolidation. The numerical data can also be interrogated to differentiate between the material losses caused by entrainment and consolidation.

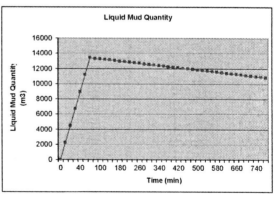

Figure 7: Quantity of Liquid mud against time

This is represented graphically in Figure 8 and it is readily apparent that consolidation is the cause of significantly greater material losses than entrainment into the water body above the fluid mud layer. The data suggests that only approximately 8% of the material is lost by entrainment and that the remainder will settle into the deeper sections of the main channel. The model's output also suggests that after approximately 60 hours all the fluid mud has either settled onto the sea bed or has been entrained into the water column.

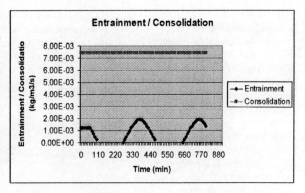

Figure 8: Relative effects of Entrainment and Consolidation

Sensitivity Analysis

From sensitivity analyses carried out on the output from the "RAT", it is known that the rate of entrainment from the fluid mud layer is highly dependent on the Bingham Yield Stress. For instance, if a value of 1Pa is assumed for this parameter (instead of the 2Pa initially inputted into the model) the rate of entrainment becomes of a similar magnitude to that for consolidation.

Similarly, the relative effects of entrainment and consolidation are sensitive to the grain size of the material within the fluid mud. If the material within the fluid mud is assumed coarser, the higher becomes the dewatering velocity and as a consequence consolidation increases. This therefore makes entrainment less significant in the fluid layer losses.

As part of a sensitivity analysis of the output from the "RAT" for Custom's House Quay, additional modelling was undertaken to reflect a change in the material particle size and hence the dewatering velocity. The dewatering velocity in the original modelling was assumed as 5×10^{-6} m/s, which is the settling velocity of 2µm particles in a fluid of concentration 0.05 (i.e. a density of 1100kg/m^3).

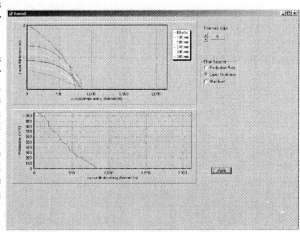

Figure 9: "RAT" Output for Revised Dewatering Velocity

The secondary analysis used 6×10^{-5} m/s as the dewatering velocity with all other parameters remaining constant.

The output from the "RAT" with this revised input parameter is shown in Figure 9.
Comparison of Figures 6 and 9 clearly demonstrates a significant reduction in the maximum transport distance achievable with the assumed larger material particle size / increased dewatering velocity. The maximum transport distance for the larger particle size material is now in the region of 850m (compared to the initial case of approximately 1200m).

In reality, greater transportation distances would normally be expected than those predicted by the model for both instances. The distances predicted have been limited due to the main channel of the River Thames having been modelled as being flat for simplicity.

The comparison of Figures 6 and 9 also shows significant differences in the fluid mud layer thicknesses at any given time. For example, after 6 hours the mud fluid layer thickness in the first analysis was approximately 1m. In the secondary analysis, with the larger assumed particle size, the equivalent thickness is less than 0.5m. This confirms the expectation of the mud fluid layer settling faster due to the increases in material particle size and dewatering velocity. It should also be noted that material losses in the fluid mud layer due to entrainment are significantly reduced in the second example. If, in any particular application, it is necessary to precisely determine the ratio between fluid mud losses caused by entrainment and consolidation, it is essential that detailed information on the rheological properties is available, in particular the Bingham yield stress and dewatering velocity.

From the lower graph in Figure 9, an equivalent in-situ production rate can be calculated in the same manner as from Figure 6. This calculation produces an equivalent in-site production rate of approximately $750 m^3/hr$, significantly less than from the output in the Figure 6 (approximately $1,200 m^3/hr$).

Conclusions

In general, Hydrodynamic processes are relatively cheap and easy to use. However, a potential problem with these techniques has previously existed that, once mobilised, the control of the sediment was limited and its final destination difficult to predict.

Van Oord, in co-operation with various Clients and Research Institutions, have undertaken extensive research to demonstrate the impacts of Water Injection Dredging on the Environment. Through this research, Van Oord and Delft Hydraulics have developed a two-dimensional model (The "Rapid Assessment Tool") to represent the Water Injection Dredging process and assess the production rates and transportation distances that can be expected in specific applications. The model is particularly suitable to studies in estuarine waters as it is capable of assessing the fluctuating velocity profiles caused by the interaction between fresh and salt water and tidal currents.

48 MAINTENANCE DREDGING II

The use of the "Rapid Assessment Tool" has been demonstrated through the undertaking of a case study on the Water Injection Dredging works regularly undertaken at the Custom's House Quay on the River Thames. The output from the model predicts production rates and transportation distances that are consistent with historical records and empirical observation.

Sensitivity analysis of the model's output has also been undertaken during the case study, with the effects of changes in rheological and physical properties of the fluid mud having been assessed and proven to be consistent with expectation.

Recent Developments in Trailing Suction Dredgers

Mr. S. Vandycke, Dredging International, Zwijndrecht, Belgium

Abstract
The design and constructing of a trailing suction hopper dredger - TSHD - has always been a special event in which all conventional ship design parameters were rearranged. Compared to a commercial vessel, a TSHD was a very costly vessel for its size. One of the most extraordinary design parameters is that the vessel will sail for 50% of the time empty. This is unthinkable for any commercial cargo vessel, but a given fact for a hopper dredge, and this fact influences the complete design of the vessel.

During the last decades there was a great evolution in the design and construction of a trailing suction dredger. This evolution was clearly visible in the size of the dredgers, but also took place in the general layout of the vessel, as well as in the ever-increasing importance of electronics and automation in the operation and control of the vessel.

Another important factor that has influenced the design of the TSHD is the environment. Over the years the dredgers did not get a good image regarding the impact and protection of the environment. The environmental issues are still becoming more and more important in the execution of a dredging operation. New vessels have special features to limit the impact of the dredging operations on the environment.

This paper will give an overview of the latest developments and evolutions in the trailer suction hopper market and will pay extra attention to the recent specific changes, requested by the evolution of the market, in the trailer suction market.

The Jumbo hoppers
The first Jumbo hopper dredge that exceeded a hopper volume of 15.000 m3 was commissioned in 1994 to Baggerwerken Decloedt. The PEARL RIVER with a hopper volume of 17.000 m3 was at that time by far the biggest hopper dredger ever built.
It was built for big land reclamation projects where a large volume of sand had to be dredged and long sailing distances were required. The large hopper volume decreased the cost per cubic meter sand dredged, as larger volumes were moved at a similar cost.

The technical know-how for the dredging installation on board of the PEARL RIVER was based on the dredging installation of the ANTIGOON that was built in 1989. The ANTIGOON had a 8.400 m3 hopper well and only one suction pipe with diameter 1200 mm. That was quite exceptional for that size of dredger at that time. The REARL RIVER had two suction pipes with each a similar dredging installation and a double hopper volume. This indicates that the technical know-how to build such a dredging vessel was available. The decision to actually build it, was a pure commercial one.

Maintenance Dredging II, Thomas Telford, London, 2005.

As we all know, this decision was a good one as the trailing suction hopper dredgers became larger and larger and were employed in the large reclamation jobs in the Far-east (mainly Singapore and Hong-Kong). Today there are trailing suction hopper dredgers with a hopper volume of 34.000 m3 and the PEARL RIVER has been enlarged up to a hopper volume of 23.000 m3.

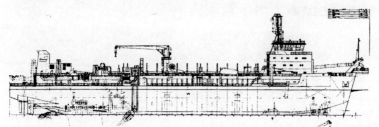

Figure 1: TSHD PEARL RIVER with 17.000 m3 hopper volume (1994)

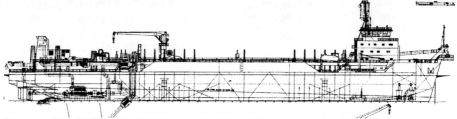

Figure 2: TSHD PEARL RIVER with 23.000 m3 hopper volume (2003)

One of the consequences of this evolution is that the latest generation of jumbo trailers has become unfit for maintenance dredging. Due to the dimensions and the increased draft, these vessels are no longer fit to work as a maintenance dredger in a busy waterway and harbour entrance channels. It becomes clear in the next table that the jumbo trailer has dimensions that no longer permit the vessel to execute maintenance in most of the cases. The UILENSPIEGEL with a hopper capacity of 13.700 m3 can still be used for maintenance dredging as the draught and main dimensions do not exceed much the ones of the ANTIGOON. The UILENSPIEGEL can be seen as a multi-functional hopper dredger that can be used for sand supply as well as for maintenance dredging in a competitive way.

Table 1: Comparison of dimensions

	PEARL RIVER	UILENSPIEGEL	ANTIGOON
Hopper volume	24.146 m3	13.700 m3	8.400 m3
Max. draught	10,8 m	9,8 m	8,7 m
Draught empty	5,2 m	5,0 m	4,2 m
Length overall	182 m	143 m	115 m

Shallow dumping
Especially for maintenance dredging on rivers and inland waterways, the draught of the vessels has to be limited as most of the dumping grounds are very shallow. It is a general evolution to limit the number of dumping grounds to limit the environmental impact by the dumping. The procedures to apply for a dumping ground have also become much stricter so the number of location that are fit to dump dredged soil are reduced. This requirement has an impact on the design of the hopper dredge as the hoppers have to dump shallower to optimise the use of a dumping ground.

In order to be able to dump in shallow dumping grounds without damaging the dumping doors, there was the development of different dumping doors and systems. It is very important that no parts of the doors come below the bottom of the vessel during dumping. Several dumping systems have been designed for that.

Sliding doors
This type of doors does not rotate around hinges, but slide sideways thus opening and dumping. This system has as major advantage that no parts of the doors can be damaged during dumping, but the system has some disadvantages:
- The total surface of all doors is limited, as the door has to slide away in an area, which cannot be used as door.
- The doors can only be opened simultaneously. One or two hydraulic pistons manipulate all doors. They cannot be opened separately.
- The seals of the doors are most of the time pneumatic. The seals are depressurised before the doors are opened, this prevents damage during dumping. The pneumatic seals remain a fragile part of the dumping system and in case of damage the vessel must be dry-docked for repair.

The system with sliding doors is no longer used in newly built hopper dredgers.

Split hoppers
The split hoppers have no doors but the hull is constructed out of two separate parts. When the hopper has to dump, the two parts are pushed open and the cargo falls out. This is the most spectacular dumping system as it looks like the vessel in splitting in two.

The split hopper design also has a number of disadvantages:
- The complete construction is rather complicated and makes it relatively expensive.
- It can only be used for small trailers as the construction becomes to heavy for bigger hopper volumes. The biggest split hopper ever built is the VLAANDEREN XX and has a hopper volume of 5.000 m3.

On the other hand, the split hopper has a very fast dumping time. The area of the dumping doors related to the hopper volume is quite high. This is also an advantage when dumping sticky soils like soft clay. The rest load in the hopper is minimal when using a split hopper.

Today, split hoppers are no longer built as the construction is complicated and to expensive compared to conventional dumping systems. Self-propelled dump barges still are built this way, as their hopper volume is relatively small.

52 MAINTENANCE DREDGING II

Figure 3: Split hopper ORWELL during dumping

Pre-dumping doors
The latest development in shallow dumping is the use of pre-dumping doors. The trailer has conventional dumping doors, but also has a number op pre-dumping doors. These are doors that are installed in the bottom of the vessel but are constructed in the central keelson tank. This way the doors remain above the bottom of the vessel even when opened. This construction is shown in the figure below.

The dumping procedure is as follows. The vessel positions itself on the dumping ground, the pre-dumping doors are opened, and the vessel starts dumping. When there is sufficient water between the bottom of the vessel and the waterbottom, the other bottom doors are opened.

With this system it is possible to ground the vessel, open the pre-dumping doors for unloading, then open the bottom doors and sail back empty to the dredging area.

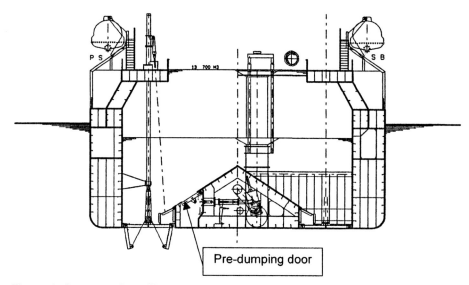

Figure 4: Cross-section of hopper with pre-dumping door and conventional bottom door

Dimensions

The design dimensions of trailing suction hopper dredgers have changed over the years. New design tools, like numeric modelling have made it possible to design the construction of a dredger in a more efficient way. Some new features have been introduced in the construction of hopper dredgers that were never seen before.

Overall dimensions

In the search for vessels with large capacity and a minimum draught, the complete shape of the hull has been modified. The vessels have become wider in order to reduce the draught as much as possible. Due to this, the propeller-shafts have a large intermediate distance and that generates a very manoeuvrable vessel.

In the table below we made the comparison of two vessels of a different generation. The JADE RIVER was built in 1978 and was, and still is, employed most of the time for maintenance dredging on the river Scheldt. The PALLIETER is the most recent vessel in the DEME-fleet and is built as a replacement for the JADE RIVER.

Table 2: Comparison between existing and new hopper dredger

	JADE RIVER	PALLIETER	Difference
Year of construction	1978	2004	
Length overall	100 m	98,2 m	-2%
Width overall	16,4 m	21,6 m	+35%
Hopper volume	3.280 m3	5.250 m3	+60%
Loading capacity	5.640 Ton	8.100 Ton	+44%
Total dieselpower	6.222 kW	6.776 kW	+9%
Draught	7,3 m	7,1 m	-3%

In this table it is clear that a recently built vessel with similar dimension outperforms an existing vessel by far.

Diesel engines and engine room

The latest development in the diesel engines, used as main engines, in a hopper dredger show us that the dimensions of the overall dimensions of the engine for a certain power output has reduced drastically. This means that the volume taken in by the engine room on a hopper dredge can be reduced, giving the designer more space for the hopper.

The new un-manned engine rooms in the vessels also create more space, as a separate control room is no longer necessary. This also allows the designer to increase the hopper volume.

Bulb

The bulb has been a very common feature on commercial cargo vessels for quite a long time. The problem with hopper dredgers has always been that a hopper dredger sails empty for half of the time. A conventional bulb is designed for a loaded sailing condition only. When these conditions are not met, the bulb can have a negative effect on the sailing speed and the fuel consumption.

The bulb for a hopper dredger is designed for both sailing conditions, loaded and empty, and has several functions:
- To decrease the resistance of the hull in the water while sailing. This results in a higher sailing speed and/or reduced fuel consumption.
- To influence the trim of the vessel so that the vessel has a better trim while loaded. This results in a more efficient hopper load and a higher cycle production.

The LANGE WAPPER was built in 1999 and was designed without a bulb. When the sister ship UILENSPIEGEL was built in 2002 it was decided to modify the design and integrate a bulb on the vessel. This had the following impact on the performance of the vessel:
- Sailing speed: increased from 14,2 knots to 15,7 knots
- Effective load in hopper increased with 5 - 10%.
- The waves generated during sailing are much less. This is clearly to be seen on the pictures below

Figure 5: Comparison of a vessel with and without bulb

Environmental aspects
A trailing suction hopper dredger has never had a good reputation when the environment is concerned.

Turbidity
Turbidity we define as "The particles that come into suspensions as a result of the dredging operation but are not removed". Even when the soil is not contaminated, this turbidity can create a cloud of particles in the water and this can have an influence on the sea-life in and around the dredging area.

The turbidity created by a hopper dredge has two main reasons:
- Overflow
- Lean mixture overboard (LMOB)

When soft soils are being dredged, the soil does not settle in the hopper so the operations are performed without any overflow. In this case it is important that the soft soil is being dredged at high density. This can be achieved by a good degassing installation and a well performing pumping system. The latest hopper dredgers can dredge soft soil at an in-situ density. This reduces the total volume of soil to be dumped and improves the efficiency of the dredger. The turbidity created by the draghead itself is minimal.

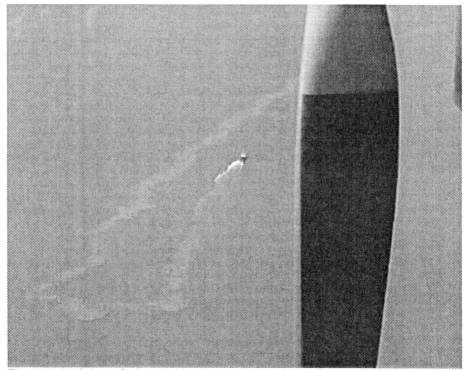
Figure 6: Arial view of a hopper dredger working with overflow and LMOB

The table below shows that when the dredger works without LMOB and overflow, the turbidity is reduced to an acceptable level.

Table 3: Turbidity generated by a hopper dredge

	Production rate [m3/OH]	Increase in suspended solids (C-factor) [mg/l]
Trailing suction hopper dredger with LMOB	5.500	400
Trailing suction hopper dredger without LMOB	5.400	150

However when sand is dredged, we want to limit the turbidity created by the overflow. Therefore some special design features have been introduced in the latest hopperdredgers.

Low-density hopper

In order to reduce the number of particles that flow away from the hopper through the overflow there is only one option. That is to increase the time the particle is in the hopper and give it more time to settle. This means that the volume of the hopper becomes larger for the same maximum load.

This has an impact on the general construction and lay-out of the vessel as the same strength has to be provided with less material. A large part of the cross-section has been made available as hopper. The figure below shows the differences in cross-section from a recent hopper dredger, the PALLIETER and its predecessor the JADE RIVER.

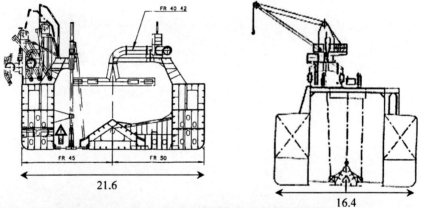

Figure 7: Typical cross-section of PALLIETER and JADE RIVER

Table 4: Hopperdensity for PALLIETER and JADE RIVER

	PALLIETER	JADE RIVER
Hopper volume	5.250 m3	3.280 m3
Loading capacity	8.100 Ton	5.640 m3
Hopperdensity	1,54 Ton/m3	1,72 Ton/m3

Overflow and loader constructions

In order to minimize the overflow losses, a number of studies and research projects have been executed in order to determine the best overflow construction and the best position of the overflow in the hopper. On the other hand studies and tests have been executed to determine the best shape of the loading installation in the hopper. The water-sand mixture that is pumped in the hopper has to be introduced in the most gentile way. The less turbulence that is created at the entrance the faster the particle will settle and thus reduce the overflow.

Older vessels have one or more chutes on which the dredge material is pumped and is distributed in the hopper. This chute is above the waterlevel in the hopper at all times. The material that falls in the hopper creates a lot of turbidity in the hopper. This means that the settling time for a particle is increased.

The new vessels have a so-called 'deeploader' instead of a chute. The sand-water mixture is spread as good as possible over the full width of the hopper. The deeploader is constructed to minimize the energy with which the mixture enters the hopper. The construction assures an even spreading of the sand over the total width of the hopper, as the construction has practically the same width as the hopper construction itself.

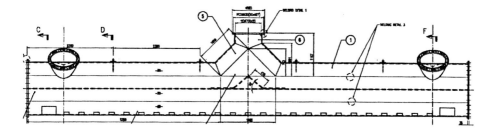

Figure 8: Typical construction of a deeploader of a hopper dredger

The energy in the mixture is reduced in two levels. In the first level the mixture is spread over the total width of the hopper and in the bottom of the first level there are openings that allow the mixture to flow down at a reduced flow rate to the second level. From the second level the mixture enters the hopper at a reduced flow rate with a minimum of turbulence.

Conclusions

The recent developments in the design and construction of a trailing suction hopper dredger has several reasons but they can be divided in two main sections:
- The requirements of the market.
- The technical evolution in design and construction techniques.

In the first section, the environmental issues generate the main influence as more and more importance is given to the environment and the impact on the environment by the dredging operations. The overall design of the dredger is influenced by this requirement in order to minimize the environmental impact of the dredging operation.

The second section is mainly a consequence of the ever-present development of techniques. The improvement of the performance of the diesel engine also has an impact on the overall design of a vessel. The ever-increasing importance of automation in all industries also improves the performance of the dredging vessels.

On the other hand the new design tools like numerical modeling have their influence on the construction of the vessel. With these computer programs the steel used to build a vessel is used in a more efficient way. In general less steel is used and sometimes in other locations then previously.

The result is that the trailing suction hopper dredger is becoming more efficient and more versatile. The vessel is more and more used for other purposed than maintenance dredging.

Mechanical Dredgers for maintenance and contaminated sediments

Sander Vlug, Baggermaatschappij Boskalis b.v, Papendrecht The Netherlands

How it all began

Maintenance Dredging II, Thomas Telford, London, 2005.

Why utilise mechanical means?

Every project has its own characteristics; therefore within these characteristics it has to be ascertained what is the optimal methodology to execute such a project.

These characteristics can be roughly categorised in the following groups:

- Soil
- Depths
- Dredge area
- Disposal area
- Environment

Any one, or a combination of some, of the above characteristics will reduce the possible methodologies to in most cases to just 1 or 2.

General dredging methods are characterised by the main equipment utilised and can be divided in 'stationary dredgers' and 'non-stationary dredgers'. For the non-stationary dredgers you have to think of the Trailer Suction Hopper Dredge and for the Stationary dredger, The Cutter suction dredgers and the Mechanical dredgers.

Some of the reasons which in general lean towards use of mechanical dredgers are as follows:
- 'Debris' in the dredge area (wires, tyres, building debris etc. etc)
- Confined working areas (quay-walls, shallow surrounding seabed, jetties and corners)
- Shallow working area
- If we require to load barges
- In case of polluted soils or other environmental constraints.

Equipment utilised in present day

If we look at the 'mechanical' dredgers that are being used today, we see that the bucket dredger is still around together with the grab and backhoe dredger. Another piece of equipment that is sometimes utilised is the 'plough', either in combination with other dredging equipment or as primary dredging unit.

The Bucket dredger:

The Backhoe Dredger

The Grab Dredger

The Plough

Working Environments

In present day there is a lot more awareness of the environment. Dredging methods have therefore, where required, been adapted to minimise negative influence(s) on the environment.

We can distinguish different cases within the maintenance dredging sphere:

1. Dredging in 'clean' soils in environmental un-sensitive areas.
2. Dredging in 'clean' soils in environmental sensitive areas.
3. Dredging in 'polluted' soils in environmental un-sensitive areas.
4. Dredging in 'polluted' soils in environmental sensitive areas.

In a case 1 scenario conventional dredging can be utilised. In all other cases special measures are to be taken to protect the environment and in case 3 and 4 where we have to deal with polluted soils, also the excess-dredging volume should be minimised in order to minimise the environmental and cost impact.

Measures to protect the environment vary from 'physical barriers' to applied dredging techniques and special tools.

It are mainly the 'special tools' that have found their way into mechanical dredging.

Special Tools

When we refer to special tools we must think of innovations to existing concepts. Where in mechanical dredging we are using mainly scooping devices such as buckets and grabs it is in this field that the innovations have materialised.

In conquering the environmental problems one has to first identify the potential problem causing effects of the dredging process.

1. Turbidity
2. More volume, more impact

The effects of both causes can be minimised by improvement of dredging accuracy end efficiency, so the use of accurate positioning (DGPS) integrated in an advanced Crane Monitoring System (CMS) is a must.

The effects of the first cause can furthermore be minimised by using a closed grab or closable bucket.

A normal grab can be easily modified by fitting rubber flaps on top of the 'shells'
The closable bucket, 'visor bucket', is a 'light weight' bucket with a lid that is to be hydraulically closed after the bucket is filled and before the bucket is being brought up through the water column.

The effects of the 2nd cause have been minimised by the introduction of the Horizontal Profiling Grab (HPG) and the Horizontal Closing Wire Grab (HCWG). The use of these grabs, in combination with an even more advanced positioning system, such as Real Time Kinematic GPS (RTK), integrated into the Crane Monitoring System (CMS) have lead to considerable higher dredging accuracies.

The Horizontal Profiling Grab

A normal grab, once it is lowered on the seabed, pulls itself into the seabed, while closing, leaving behind a kind of egg box pattern. A HPG or HCWG is designed to close horizontally, therefore if used in combination with RTK positioning and a CMS system the seabed can be dredged more evenly therefore less excess materials have to be dredged to achieve the contractual depth throughout.

The main difference between the use of the HPG over the HCWG is that the HPG is used from a backhoe dredger with a rotor between stick and HPG and the HCWG is used from a grab dredger using a tag line. This results in the following advantages of the HPG over the HCWG:

- Superior horizontal positioning
- Dredging grid is rectangular against circular

This leads in general to higher bucketfill and less over-dredging on slopes, which in turn will result in lesser scoops, therefore lesser environmental impact.

It has to be noted that with the introduction of high accuracy positioning and accurate visualisation of the dredging process, the dredger is still manually operated and therefore the accuracy achieved is still very much related to workmanship of the operator. We have only given him the best tool available.

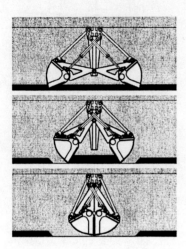

Transportation of 'dredge spoil'

The transportation of the dredged soil can be done in various ways. Traditionally it will be loaded into hopper barges alongside the dredger. If the materials have to be placed on land the material can then be placed onto a conveyor belt which transports the material on-shore or placed into a soil pump to transport it hydraulically. In case of very narrow working space it can be considered to use a conveyor belt system to load barges, for example behind the dredger.

Optimising maintenance dredging

Peter de Wit, Port of Rotterdam, Port Infrastructure, Dredging Department, Netherlands.

Introduction
In the harbours and fairways of the Port of Rotterdam maintenance dredging has to be carried out on a regular base. The question how maintenance dredging can best be done is of great importance, for the depth of the harbours and fairways must always be kept up to par.

Maintaining the required depths for shipping is one of the basic elements of port management. Minimising the costs is a major concern of the port authority.

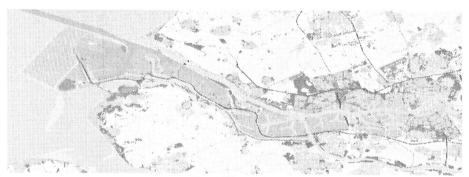
Port of Rotterdam

Maintenance dredging in Rotterdam
Due to the geographical location of the Port of Rotterdam, at the mouth of the river Rhine and on the North Sea, sediments accumulate in the harbours and fairways from two sources.
As the Rhine flows through Rotterdam to discharge its waters into the North Sea, some of the silt that it carries is deposited in the harbours and fairways of Rotterdam.
At high tide the North Sea waters enters the harbours and fairways. This water brings in silt from the sea and this also settles in the harbours and fairways.
On a yearly base in the Rotterdam situation an amount of about 4,5 million m3's has to be dredged.
Trailing suction hopper dredgers (tshd) with the assistance of grabdredgers and bottom levellers carry out almost all the maintenance dredging.

This amount is not equally spread over the year. The rate of siltation is very dependable on what "mother nature" has in mind and does not proceed according to a fixed pattern. Sometimes a great amount of sediment enters the harbours and fairways in a very short time. This means that dredging actions have to take place in short terms to restore the required depths. Additional this means that planning maintenance dredging can be a problem.

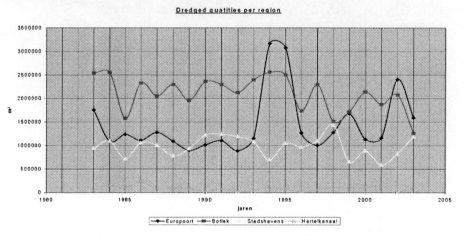

In common the depths are determined by multibeam echosoundings, with a frequency of 210 kC. In some locations of the port the depth is determined by density measurements. In these locations a density of the bottom layer of 1,2 t/m3 is accepted by the pilots. The figures are shown on survey charts. The nautical employees consider those figures as bottom.
The question is "what is bottom". In the port of Rotterdam the water bottom consists mainly of silt. Dependable on the required keel clearance it is proven that ships can easily manoeuvre through a certain mud layer. From nautical research it appears that a marine vessel is still manoeuvrable in "thick water" to a density of 1,2 t/m3.
Because of this the Port of Rotterdam is doing research on this phenomenon. Among others investigation takes place on the rheology of the silt layer.

Another subject Rotterdam has to deal with are the sailing distances for transport of the hopperload to the disposal areas. Technically dredging is not considered as a dredging problem any more but much more as a "transporting problem". Although the technical part has continuous attention, such as the hydraulic behaviour of the dredged mixture in and around the suction head, degassing of the dredge pump and the way of dredging based on density.
Also attention is given to the development of the bedlevellers by using different dimensions and accessories.
In a "birds eye view" this paper gives an overview of the several subjects on the optimising of maintenance dredging in the Rotterdam situation.

Contracting
The execution of the maintenance dredging in the ports of Rotterdam is contracted out to private dredging companies. To ensure to have dredging capacity available at any time a basic capacity is contracted. A long-term contract is arranged with a contractor, under the terms of this contract he deploys tshd's of a certain capacity per year. This basic capacity is not enough in relation to the total needed capacity, but by offering the contractor a certain continuity in

the occupation of his equipment over the year ensures him a great deal of the dredging work that can obtain economic profit.

For the remaining dredging work and by peaks in sedimentation contracts are put out on a job-by-job base by means of so called "option charter contracts". Annually all contractors are given the opportunity to indicate what they have to offer. For each case a selection can be made for the most suitable available equipment for the job which has to be carried out.

On the base of this a assignment can be made with a selected contractor for a specific part of dredging. In this way it is guaranteed that the basic of the dredging will be carried out while having the freedom to carry out the rest of the dredging as efficiently as possible.

Trailing suction hopper dredger.
The quality of the silt is yearly assessed throughout the entire port area. For this purpose bottom samples are taken and investigated. On the base of the analyses the degree of contamination is determined. Owing to the mixing of the river and sea silt a decrease in the degree of contamination from east to west can de observed. The degree of contamination determines whether the silt can be disposed at sea or must be stored on a landdisposal site. For storage of the contaminated sediment the large scale disposal site "Slufter" is available, which was built in 1987 beside the coast of the Maasvlakte.

Transportation
The dredged not contaminated silt can be dumped in the North Sea. For this in the North Sea a number of dumping locations are designated. However the at those places after discharging disposed silt does not in total remain on the seabed, but under the influence of waves and current it is partly spread over the sea bottom.

After research on this it is decided to dig pits in the sea bottom near the port entrance in which the silt can be disposed. The construction of those pits is undertaken by tshd's in combination with the executing of the maintenance dredging.

When loaded the tshd's have to sail to the dumping area at sea or to the shore delivery point so that their pumping units can deliver the silt into the disposal site. For this large sailing distances has to be covered.

The complete dredging process is defined by:
- sailing empty
- dredging
- sailing loaded
- discharging
- idle time

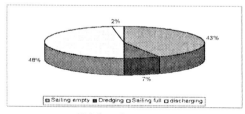

Tshd cycle

Comparing to the total cycle time of a tshd trip in maintenance dredging it is obvious that the time of dredging contains just a slight part of this cycle. The greatest part is spend on transportation.

In this the objective is to reduce transporting costs by avoiding the tshd sailing empty from the discharge areas to the dredging spot. In the Rotterdam area there is a great need of sand for infra structural projects. To supply a part of the own needs of sand use is made of the empty sailing of the tshd. In combination with the maintenance dredging cycle after discharging at sea, sand is dredged at sea, transported to the harbour and discharged in depots.

In this way part of the not-economical empty sailing time is of advantage. Sand not needed for own projects is brought into the sand market on an economical base.

Also taken in consideration is to find out whether dredging by means of agitation is possible in some not contaminated locations of the port. For this the systems of water injection or water/air injection can be very useful. On this subject investigations and tests are implemented. The objective is to make visible the behaviour of the suspended silt by means of different tracing systems.

If a system like this can be implemented it will provide a great reduction on the transport costs to the dumping areas. An additional advantage will arise by sparing the environment because of the less emissions of exhausting fumes of the dredgers and the need of less energy.

1,2-Dredging

In some parts of the ports dredging is done with the aid of so-called "density charts". In these charts the sounded depth is reduced by the distance between the values of 1,03 t/m3 (sounding) and 1,2 t/m3 (nautical density).

As the performance of the tshd is measured by the hopper volume with a 1,2 t/m3 density, silt of a density higher than 1,2 t/m3 is usually dredged. For the contractor this ensures that a 100% hopper is dredged and that the process will run evenly, which shortens the suction time. In areas with a thick layer of silt, the dredge master lowers the suction head into the siltlayer to a depth that will ensure the most optimal running of the dredging process. This means that the suction head is put in the lower part of the 1,2 t/m3-layer and often deeper.

During intensive dredging actions it appears that mixing of the 1,2 and 1,3 layers takes place. The result of which is a homogeneous layer with lower specific gravity. The 1,2 level, however, remains at the same depth as before the dredging action.

This is why the high hopper performances cannot be recognised on the sounding and density charts.

The process described above has become known as the "fluffing-up-effect". In the short term it seems to exert more influence on the bottom position (1,2) than the consolidation process. A dredging method was selected in which the silt layer at the top of the 1,2 t/m3 layer would be removed. If the suctionhead is not lowered into the silt layer with higher density, but kept high in the 1,2-layer, the silt is removed from the place the sounding chart refers to. ("You've got to remove it where it hurts").

By this approach it is achieved that the density structure of the silt layer is transferred in thinner layers in raising of the 1,3 level. As a result of this the in-situ-result is more in comparison with the result measured in the hopper.

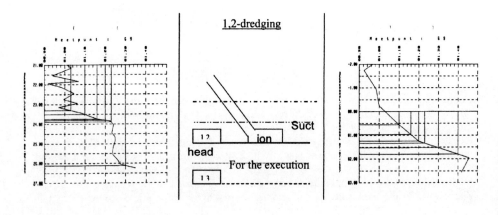

of maintenance dredging in the ports of Rotterdam the tshd's mainly are equipped with the California suction head. The central question is "what is the hydraulic behaviour of the silt in and around the suction head during the dredging process and how can it positively be influenced".

Effort has been made to calculate the forces between the silt and the underpressure in the suction head. Also to get any insight in what is happening in and around the suction head the course of the pressure was measured, this in combination with the position of the visor.

The visors are adjusted in a way by which an opening arises perpendicular to the sailing direction. Due to this the silt will be forced into the suction head, through which less under pressure in the suction head is needed. It seemed that where the amount of silt is the greatest, the highest pressures are measured. In the middle of visors the lowest pressures are measured. The pressures do increase towards the edges of the visors.

Out of scale models it is observed that when too much silt is forced into the visors, this will have a negative effect on the yield of the dredge pump.

By means of the shape of the suction head and the position of the visors a "dead angle" in the suction head will arise in which the silt will accumulate. Owing to this the effective surface of the suction head is reduced. The surplus of silt is pushed towards the surrounding of the suction head. This can cause a turbulent flow and fluffing up of the surrounding siltlayer.

Continuation of the research on the interplay of the mentioned forces and pressures in combination with the functioning of the different types of suction heads will be undertaken.

Bottom leveller

In principle the bed leveller is used to replace the silt from locations which can not economically be reached by a tshd. Such as in front of quay-walls, jetties and spots at the end of a harbour.

When a tshd-action has taken place it is possible that rigs will arise in the silt layer. It is inefficient to remove those rigs with a tshd. To avoid overdredging by the tshd a bed leveller can flatten out the rigs by replacing them into the deeper parts. By doing so "chasing rigs" with a tshd can be avoided and extra tshd work can be prevented.

For this purpose the bed leveller working in the ports of Rotterdam is also provided with a different type of blade which can be compared with a bulldozer blade.

Next to a tshd action with the bedleveller an "after care" action can be executed to reduce the top of the rigs and to recover the harbour bottom. As a result of this additional tshd work is not necessary.

Water bottom

As the harbour beds in the ports of Rotterdam consist mainly of thick silt layers it is not easy to determine the "harbour bottom". Concerning the needed keel clearance for the sea going vessels there are benefits that can be gained from the adoption of the nautical depth approach. This approach should help to reduce the quantities of silt, which require dredging. In the port of Emden in Germany the nautical depth is defined by shear parameters. In this port one has managed to develop a technique for measuring the in-situ yield stress and can determine the nautical depth based on a yield stress profile rather than a density profile. For this may be a better way in defining the nautical depth since it is the internal shear strength of the mud which affects navigation through it.

By means of research and measurements in co-operation with the port of Emden Rotterdam has started to investigate the possibilities of translating the Emden experience to their own situation. This to find out whether a navigable siltlayer, different from the current approach, can be detected and introduced.

Supervision

Instrumentation plays an important role in dredging. Formerly more simple "instruments" were used, for instance a paint mark on the wire to indicate the depth of the suction head and above all the experiences of the crew ("the eye of the dredgemaster"). But a greater variety of instruments and automation found their way to the dredging operations. An overwhelming amount of information is available.

For the supervision on the daily dredging practices in former days on board the dredging vessels two inspectors were on duty. One at day and one at night shift.

Nowadays all dredging operations are supervised from a shore station, called the "dredging desk". All technical dredging data in combination with the positioning of the dredging equipment are on line presented on displays.

The system must make visible:
- where (position)
- what (process data)
- quantity (time, etc.)

By means of this presentation of data in co-operation with the contractor the dredging practices are carried out. The collection if this data provides also the opportunity to evaluate the works and for the use of innovations.

During the complete time of operation a presentation is given of the location of the dredger in x and y and in which direction the dredger is progressing (course).

In the dredging phase all relevant process data and the survey information (dredging spots) complement these data. The survey information is simultaneously presented.

Renewing dredge data from surveys can also be done more accurate and quicker, this can achieve a greater efficiency.

Another option leading to greater efficiency is that the sailed track is presented, there fore more effectively corrections can be made. Especially during spot-dredging this proved to be very useful.

Dredging desk

Conclusion

In a quick scan are shown the basics in the innovation of the maintenance dredging in the port of Rotterdam. The results achieved so far are only possible in a good working relation between the employer and the contractor. As an employer we must avoid to consider the contractor just as a "rent-a-hopper-firm".

To optimise the dredging practices it is also of importance to ask one self:
- Is my way of executing efficient enough?
- Can it be done in another way?
- How does my neighbour do it?
- Is his situation comparable with mine?

In other words: "look for reference points to test your own methods of dredging".

Dredging of Barton Broad; Problems Faced and Lessons Learnt

Trudi Wakelin Broads Authority, Norwich, UK

The Norfolk and Suffolk Broads

The Broads area extends over the Norfolk and Suffolk border, on the East Anglian coastline and consists of 125 miles of interconnecting rivers, with shallow Broads created from medieval peat diggings.

The executive area is managed by the Broads Authority, which is a Special Statutory Authority with the same duties as a National Park, along with the additional advantage of being responsible for navigation. The Broads Authority is the third largest inland navigation authority after British Waterways and the Environment Agency. Additionally, the Authority is, in common with the other national parks, a District Planning Authority.

Barton Broad, Fig 1, extends over 77ha and is the second largest of the Broads in the system.

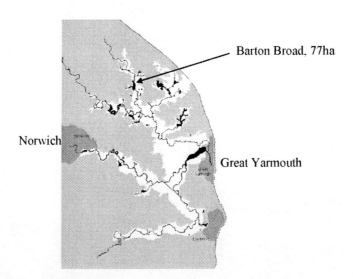

Figure 1

Clear Water 2000 Project

The dredging of Barton Broad was completed as part of the £3.3m Clearwater 2000 project, launched in 1995 with the objective of restoring the Broad to a healthy and attractive state, to benefit wildlife and increase public enjoyment. The aims included; firstly, to remove 300,000m3 of phosphate enriched mud to improve water quality in the Broad, then to undertake biomanipulation of the Broad to restore the ecosystem and to restore Pleasure Hill Island, which was badly eroded.

Additionally included in the project was to build a freshwater ecology study centre, for environmental education, also build 610m of interpretative boardwalk to provide land based public access and further educational opportunities through alder Carr woodland habitat, also including a viewing platform to the biomanipulation area and vistas over the water. Finally, the provision of environmentally friendly boating access was included, and this was achieved through an interpretive boat trip on the UK's first solar powered passenger vessel, 'Ra'.

The project was shaped not only by the Broads Authority's duties and remit but was tailored to suit grant aiding bodies such as the Millennium Commission, as without significant external funding the Broads Authority would not have been in a position to undertake the works. Agencies such as Anglian Water and the Environment Agency have also been practically involved through introducing phosphate-stripping measures to upstream treatment works and also with scientific research projects and monitoring.

Other partners included English Nature, Countryside Agency, East of England Development Agency, Norfolk Environmental Waste Services and the Soap and Detergent Industry Environmental Trust.

Project elements

Focussing specifically on the water-based works the project can be broken down into these elements:
1. The dredging contract, for which the Broads Authority employed a consultant to prepare the contract documentation and interview tenderers, and adopted suction dredging as the most suitable technique for the majority of the works.
2. The survey works.
3. Pre-contract trials.
4. Earthworks and drainage element in construction of the settlement lagoons.
5. The restoration of the eroded Pleasure Hill Island, the only remaining island in the Broad, through piling and backfill.
6. Biomanipulation.

Dredging contract

It has already been mentioned that the contract was to remove 300,000m3 of silt by suction dredging, with this material being pumped at 90% water content. Floating pipelines were used, although these had to be submerged across navigation channels, and the working areas were buoyed and roped off to prevent access thereby reducing the risk of vessels overrunning the guide wires or anchors.

The dredging costs approximated £7/m3, including considerable land compensation fees, consultants and ancillary works. The specification varied widely across the Broad, and the final contract period was 6 years.

Dredging specification

When determining the appropriate dredge level it was necessary to consider the conservation requirements to remove all silts with phosphorous concentrations above 500mg/kg.

However, in some parts of the Broad a greater depth was required for navigation but this need had to be balanced against the importance of not disturbing or exposing the peat layer, which would otherwise increase phosphorous release.

Therefore, dredging into or just above the marl layer was the optimum level. The final analysis and assessment of these surveys resulted in the wide variety within the contract specification, balancing the needs of conservation, navigation, practicality and economy.

As one would expect the deeper water can be found in the central navigation channels, and shallows out to an undredged margin around the Broad, where littoral reed swamp margin was encouraged to extend its range. Once this exercise was complete it was possible to calculate the volumes of silt arising from the project.

Hydrographic survey

An independent hydrographic surveying company was contracted to undertake the initial bathymetric survey, which was used in conjunction with cores to determine the bed and substrata profiles. The survey, Fig 2, shows the concentration of transducer soundings, which were crosschecked against lead line levels, and reduced to Ordnance Datum with reference to a Bench Mark at Barton Turf Staithe. Interim and final surveys were also carried out to monitor compliance with the dredging specification as well as calculate volumes for contract payment.

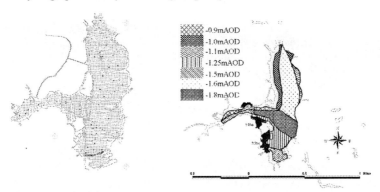

Figure 2

Pre contract trials

In 1995 a trials were undertaken to check the feasibility of the works including:
- Feasibility of pumping distances.
- Appropriate percentage solids for pumping.
- Lagoon design and water loss.
- Settlement rates for volume calculation.
- Detrimental affects to arable land.

These trials confirmed that pumping up to 1200m would be achievable at a 90% water content discharge.

Settlement lagoons
The design principle behind the lagoons was that they should become self-sealing following some initial seepage, with the silt fines thereafter creating an impermeable layer. Water loss would then be managed via the drainage system, or via evaporation.

The trials confirmed that construction of the bund walls using topsoil from the site strip could produce an effective terracing system with minimal seepage and that little water loss was observed from the bed of the lagoon.

Laboratory tests were also carried out in tandem to look at expected settlement rates, with the information used to inform the management of the lagoons in assessing the retention period required in each lagoon, as well as calculating the volume capacity of each lagoon.

ADAS were retained to look at the soil structure of the arable land and confirm for the Environment Agency's approvals process that works would confer benefit to agriculture, and were later used to advise throughout the restoration of arable areas and report on short and long term impacts.

Ongoing monitoring was incorporated by the installation of piezometers around the perimeter of the fields to record ground water levels. Thereafter, lagoons were constructed covering over 30ha and were gravity drained, managed by a system of sluiced culverts at 3 levels set into the cross walls.

Additionally, the outfall dyke was dammed to control the discharge of return water thus ensuring the discharge consent conditions as set down by the Environment Agency were maintained.

Pleasure Hill Island restoration
Barton Broad was excavated for fuel, and as late as the 1950's it was possible to see the remnants of peat baulks which had formed embankments within the diggings. Now only Pleasure Hill Island remains, and this once formed part of the Parish boundary between Irstead/Neatishead to the south and Barton/Catfield to the north. As such it provided archaeological interest to the project as well as additional wildlife habitat and protection to adjacent littoral margins by reducing the overall fetch.

Biomanipulation
In the mid – late 20th Century Barton Broad had declined to a eutrophic state, heavily nutrient enriched with algal dominance which resulted in the loss of almost all aquatic plant life and fringing reed swamp.

Earlier research in smaller discreet water bodies had proved that through engineering the environment –'biomanipulation' it was possible to regain a complete and fully functioning ecosystem.

Whilst dredging is required to remove the nutrients that act as a fertilizer to algal growth, this on its own is not enough to regain clear water. Our research has shown that additionally fish removal is required to allow Daphnia to function. This water flea filters algae from the water column, and with a large enough population can filter sufficient quantities to clear the water.

It is necessary however to remove fish, the fleas predator, in order to sustain these conditions long enough for the return of macrophytes, which then provide shelter and habitat for these and other invertebrates. Finally, fish can be reintroduced but again controlled through the management of fish stocks to include predators such as perch and pike to control the populations of roach and bream, which have predominated.

Kingfishers, grebe, bittern and otter are at the top of the food chain and so also benefit from the improved visibility that clear water provides for their hunting.

To carry out biomanipulation, barriers (Fig 3) have been installed in a number of locations around the Broad and the waters within were electro fished, using an electric current to stun the fish before netting, and then released elsewhere in the Broads system where they could do no harm. These barriers had the additional benefit of providing shelter from wind and boat generated waves, which helped to reduce suspended solids in the water column also.

Construction of fish proof biomanipulation barrier

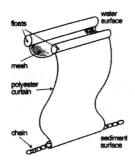

Figure 3

Within the barriers further experiments were carried out, including the installation of mesocosms to improve fish exclusion, although that turned out to be counterproductive as anoxic conditions led to toxic algal blooms therefore demonstrating the need for a minimum area to be successful. Also an experiment was undertaken with 14,000 cobweb brushes installed to provide artificial habitat for water flea and invertebrates.

For the contract period a Research Technician was employed, looking at the implementation of biomanipulation as well as carrying out a range of monitoring duties looking at dredge depth and substrate, phosphate release, suspended solids, recolonisation of aquatic and emergent vegetation and restoration of settlement lagoons for arable use (in association with ADAS).

Contract problems
No contract is without its problems, and Barton was no exception.

It is possible to broadly categorise these into Administrative; including:
1. Complexity of approvals process.
2. Stakeholder consultations.
3. Funding.
4. Lack of continuity.

And Technical, including:
i. Volume of silt.
ii. Disposal opportunities.
iii. Flooding.
iv. Biomanipulation.

Approvals process
The lead in time to the project was significant, due in part to the approvals process, although this was not confined to pre-contract, as some approvals were also required separately for each of the lagoon areas, which altered throughout the contract period. Approvals were required under the following areas; Waste Management Licencing Regulations, Habitats Regulations, Planning permission, MAFF derogation, landowner consent, and Environment Agency's discharge consent.

Stakeholder consultations
Stakeholder consultations were also ongoing. The Barton Broad Liaison Group was established early in the process and included representatives from local residents, businesses, boatyards and Parish Councils, as well as English Nature, Broads Hire Boat Federation and the Norfolk Wildlife Trust.

Funding
The funding partners have already been listed, but it is interesting to note that full funding was not in place when the contract was awarded. However, it is likely that had we waited for a complete funding package we would still not have started works. Figures were constantly changing, with promises made, broken or exceeded and resulted in insecurity. The project itself became fluid, adopting areas of work to suit grant criteria, which also brought complex claim/reporting and audit procedures.

Continuity
During the life of the project all parties to the contract changed. Initially there was a change in consultants before the project was ultimately taken in house in a bid to reduce costs. The original contractor had been a joint venture, this dissolved following one of the parties ceasing to trade. Therefore, the contractual relationship with the contractor needed renegotiation. Finally, many staff changes on the client side – from project technician through to Chief Executive – during the life of the contract also meant that information and the potential for efficiencies to be made, was lost.

Silt volume
Turning now to the technical problems that the contract experienced, the first of these was the volume of silt. It is clear that in the early stages of the contract a serious miscalculation occurred which left the project short of disposal area, either by overestimating the

consolidation of the drying silt, failure to take into account bund wall volumes and margins in the lagoon capacity or failure to recognise that the Broad was still accumulating silt at a rate in excess of 1cm per year throughout the contract period.

Disposal opportunities

The disposal opportunities were limited due to various restrictions; finance – although any problem can be overcome with enough resources; but without that option it was necessary to look for arable land, on a slope, within practical pumping distance, outside the Special Area for Conservation boundary, with amenable landowner and neighbours, and access for plant.

Having found a site that fulfilled all the above criteria, except for the neighbours, we were then beaten by geology. On first glance this site was ideal, with the only downside being adjacent properties, but these were at the top of the slope of the field and so unlikely to be anything but marginally inconvenienced. However, further investigation showed that the proposed site was an area of natural springs, where the Norwich Crag aquifer outcropped. By sealing the area with lagoons there was the potential to elevate the water table upstream and so this option was abandoned.

Flooding

In 1997, following a very wet spring period, a claim was received from an adjacent property claiming that they had flooded due to the proximity of the lagoons and their affect on the water table. Despite being able to demonstrate the water table relationship with the Broad level and the negligible effect of the lagoons this claim is still being pursued, and vigorously defended. Therefore, despite being content that the works were not to blame, future schemes now avoid properties within a greater margin.

Project results

The results of dredging Barton Broad can be summarised as follows:
- 43cm deeper on average.
- There is now 44% more navigation area accessible for sailing.
- 50 tonnes of phosphorous has been removed, which is the equivalent of 20 years loading.
- There is now 50% lower soluble phosphorous release from the dredged sediment.
- After dredging the exposed substrate takes on a similar consistency to the undredged surface sediment.
- Clear water has been attained in the biomanipulation areas, with 12 species of plant now present.
- There has been no evidence of any change in the trophic structure resulting from nutrient removal, which may be due to the resilience of fish populations that result from connection to the wider river and lake network of the Broads.
- Slow response – shallow lakes do respond to nutrient reduction over a time scale of approximately 15 – 20 years.

Lessons learnt

From these experiences it is possible to draw out the following lessons:

i. Suction dredging was highly effective, controllable and sensitive. The post dredge core survey demonstrated compliance with the dredge depth specification, and it is important that this specification takes into account the substrate exposed when considering phosphate release as well as macrophyte establishment.

ii. However, the possible accuracy of the bathymetric survey of within 100mm combined with resiltation and bed heave resulting from the removal of overburden from the peat made the method of measurement unsatisfactory for precise volume calculations.

iii. Additionally, the relationship between client and consultant is very important to fully appreciate the pros and cons of the different options of Forms of Contract available, and the client needs a good understanding of the possibility of increasing costs in a long term contract through variations in inflation rates, fuel costs etc.

iv. In planning a dredging project environmental factors need to be considered, particularly where the project objective is to result in a particular habitat or ecosystem structure. Dredging will remove shellfish so it is important to consider recolonisation – dredging over a long period such as at Barton means that mussels were able to recolonise effectively, but in other sites we have translocated rare species.

v. Adjacent plant beds or rare plants may require protection from silt curtains and palaeoecology studies should be done to look at target dredge depth if macrophyte establishment from a seed bank is a desired outcome.

vi. The light climate should also be considered as deeper = darker and it is important to consider this effect on macrophyte establishment.

vii. The timing of dredging activity is important as it can have an effect on the remobilisation of solutes; SPA wintering birds, and also on recreational activity therefore it may be necessary to try and work in restricted areas during winter months when there is less boat traffic, or concentrate on favoured bird roosts during the summer.

viii. The effect on residence time should be considered as increasing the volume of a water body will mean that water will remain in the water body for longer. This can be important for algal populations as it reduces flushing effects.

ix. As already stated, it is best to try and avoid adjacent properties as far as reasonably possible at disposal areas.

x. And finally, keep stakeholders as well informed as possible concerning the project to avoid last minute complications arising.

Our understanding in respect of ecosystem management has also improved, learning that the project was unable to effect the 'switch' to clear water conditions for the whole Broad simply by dredging but it is expected that the phosphorous reduction will allow for more stable recovery in the longer term. The practical biomanipulation techniques have also been improved through refinement of designs and working practices, and can demonstrate successfully the theoretical principles.

The effect of dredging on sediment phosphorous over time can be summarised as follows:
- Immediately after dredging the phosphorous concentration of the newly exposed sediment was reduced by 50%.

- One month after dredging the surface sediment (1cm depth) re-equilibrated to have a phosphorous concentration similar to that of the undredged sediment, resulting from sedimentation.

- Up to 5 years after dredging the sediment had significantly lower phosphorous concentration from 3 – 20cm depth.

Summary

In summary, the Barton Broad experience has shown that dredging can remove huge amounts of phosphorous. However, we face serious disposal challenges but if we could dispose of this material to arable land this could mean a net reduction in catchment nutrients although legislation currently works against this.

Ultimately a holistic management strategy is required to provide a sustainable solution to nutrient loading, which would ensure inputs were reduced still further and this may eventually be forthcoming through the implementation of the Water Framework Directive. Finally, we need to be patient; ecosystem restoration is a long-term goal.

Conclusions

Be bold – innovative projects need vision and ambition, along with the political will to carry them forward.

A detailed level of forward planning is necessary, considering the project aims and thoroughly researching all possible options in advance of final design selection. Whilst the use of consultants can assist with this process, the relationship between Engineer and Client is vital in building an understanding of the project to identify the optimum solution.

An inclusive approach to schemes in terms of statutory regulators can ease projects through approvals and permitting processes, particularly when engaging in dialogue at a very early stage. This approach will highlight any areas of possible conflict and allow the project design to be shaped by these concerns, mitigating impacts or designing out unacceptable solutions.

Likewise, engaging early with stakeholders will also allow consideration of additional concerns in advance of the award of contracts. This should avoid last minute objections or complications which could lead to delays on the scheme, thus incurring claims or other cost implications.

Don't fear failure when 'engineering the environment' – even a set back teaches a lesson and will add to our overall understanding of the complex processes that govern waterway ecology. At the same time it should be remembered that whilst general principles can be established, site conditions will also vary and can have different levels of impact therefore lessons are not completely transferable between projects and should always come with a 'health warning'.

It is important to accept a level of risk within a project where innovation is being applied, and whilst there is usually an inverse relationship between cost and risk, this general rule of thumb cannot be applied to biomanipulation.

Adequate contingencies built into a project, in terms of time as well as finance, will give comfort to the Client, and any grant awarding bodies who may be supporting the scheme, when reviewing project progress and auditing the works.

Finally, look for the 'win - win'. A multi disciplinary approach is rewarded by greater access to funds as well as better protecting the environment and also giving opportunities to people to enjoy it.

The Barton Broad Clearwater 2000 project has just won the Broads Authority Beacon Council status in the sustainable tourism sector, which will allow us to further disseminate our experiences.

Dredging and Disposal Methods for Small Projects (marinas, canals, lakes and reservoirs)

Simon Bamford, Operations Director, Land and Water Remediation Ltd

The constraints, both practical and legislative, of dredging and disposal methods in inland waterways have led to significant developments in plant and sediment treatment and disposal methods in the past 10 years.

Given the fact that inland projects can vary greatly in size, location and characteristics, the need for flexibility and particularly ease of mobilisation has led to the development of highly mobile, road transportable plant designed to work in a wide range of water environments and locations. Furthermore, the increase in Health and Safety and Environmental standards and clients drive to ever lower cost has greatly influenced plant development.

Small inland dredging projects can range in size from $100m^3$ to be removed from silt traps for urban drainage systems up to 50 - 60,000m^3 for a major canal or river scheme. In between are a plethora of streams, lakes, ponds, marinas and small rivers all requiring essential maintenance dredging but each with its own peculiarities.

The most common issues related to inland dredging schemes are:

- Access
    - restricted room for launching floating plant;
    - trees and vegetation;
    - narrow and or shallow channels, structures, buried services etc.

- Water Depth
    - Depths can range from <0.5m to typically 5m or more

- Debris
    - urban water bodies typically contain large quantities of debris such as bikes, shopping trolleys, safes etc. all of which influence the dredging and disposal options;
    - screening becomes necessary where beneficial reuse is the disposal option

- Selection of dredging techniques:
    - floating;
    - excavation following dewatering;
    - backhoe
    - suction

Maintenance Dredging II, Thomas Telford, London, 2005.

- Limited land availability
    - often the operator/owner of the water body has little or no land in their ownership limiting on-site disposal options;
    - third party agreements are often required for site accommodation, offloading, treatment and disposal;
    - significant haul distances and costs can result from transferring sediments to a different location.

A significant issue of inland dredging projects that has helped to drive plant development is access. The difficulties associated with access and the broad range of project sizes have led to the need to develop easily transportable and flexible plant with the ability to put together a wide range of configurations to suit the varying site conditions.

Key plant developments in recent years have included the move from fixed waterborne dredgers to road transportable dredging pontoons that can be broken down on site to cope with variable channel widths and constraints whilst the dredger can work from both the water and the bank thereby maximising productivity and minimising cost.

Further developments include inland tugs with remote tow linkages and hydraulic drives so giving maximum manoeuvrability in tight locations and the increase in the national fleet of long reach excavators capable of working in a wide range of locations.

Inland Dredging is most commonly undertaken using backhoe dredgers and bank mounted long reach excavators as it delivers high production rates and can deal with a wide variety of materials and debris. However, suspended solids are becoming a growing issue and the use of ecograbs is expected to grow. Suction dredging is practiced but given the difficulties of disposal and the requirement to dewater after dredging this method is not commonly used.

Agitation and water injection dredging are not commonly used methods although in tidal reaches of rivers hydrodynamic techniques have been found to be successful.

Legislation has and will continue to have a significant impact upon disposal methods used in inland dredging. With the introduction of the Waste Management Regulations (1994) dredged sediments were classified as waste with the consequence that uncontrolled disposal was prohibited and a more managed approach required.

This led to dredgings disposal costs increasing significantly unless beneficial reuse of the materials could be demonstrated. Therefore during the 1990's and early 2000's much work went into understanding the beneficial properties of sediments and their use in agriculture and land restoration projects with large volumes of materials being applied directly to agricultural land adjacent to the dredging location.

However, the enactment of the Nitrates and Landfill Directives and the changes to Waste Management Regulations exemptions, have and will continue to drive changes to disposal routes with the consequence that direct disposal to agricultural land will reduce.

The effects of legislation and environmental best practice have made the selection of the dredging, treatment and disposal process a very complex and iterative process with each element having an impact upon the other in terms of operational, environmental and cost issues. In large part these difficulties can be mitigated by ensuring early contractor involvement, an approach adopted by British Waterways through their use of a Partnering Term Contract for the management and delivery of their annual maintenance-dredging programme.

To meet the legislative challenge, new treatment techniques for not only contaminated sediments, but also 'clean' materials have been and will continue to be developed.

Examples include the composting of highly organic sediments dredged from parks in London. Due to transport constraints and limited reuse options within the park, dredged sediments have been blended with shredded green waste from a London Borough's parks and then composted to produce an organic soil for beneficial reuse throughout the Borough.

A further example is the screening, separation and washing of contaminated sands and gravels from dredged sediments and the mechanical dewatering of the fine sand, silt and clay fractions. The recovered and washed sands and gravels were beneficially reused as fill behind sheet piles whilst the pressed fine fraction was reused as topsoil for a closed landfill site.

It can be seen that the legislative changes and environmental and economic pressures of last 10 years in the inland dredging sector have resulted in great changes in dredging plant and in the way that dredged sediments are disposed of.

What is certain is that the pressures for environmental improvement and efficiencies will continue and that dredging and treatment methods and plant will develop further to meet these challenges.

Possible changes include:

- Plant
    o increased flexibility;
    o biodegradable oils;
    o reduced mobilisation costs;
    o improved dredging techniques such as wider use of ecograbs/buckets

- Water Quality
    o greater control over suspended solids during dredging;
    o effects of implementation of the Water Framework Directive

- Legislation
    o Water Framework Directive – potential for tighter control but also opportunity?
    o Landfill Regulations – the full effects are yet to be felt but are likely to increase maintenance dredging costs;
    o amendments to Waste Management Regulation exemptions – changes are still to come but will make registration of exemptions more difficult.

It is clear that the particular requirements of flexibility and legislative constraints on the inland dredging sector will continue to drive innovation in the dredging, treatment and disposal of dredged sediments to deliver cost effective and sustainable solutions to the maintenance of inland waters.

Forms of contract and methods of measurement

John M. Greenhalgh, Greenhalgh Associates, UK

Synopsis
In this paper I will review how the nature of dredging and reclamation works has been addressed in the various forms of contract published by FIDIC and how, from a quantum point of view, the method of payment and measurement impact on the risk profile.

The Red Book 3rd and 4th editions of the Conditions of Contract for Civil Engineering Works published by FIDIC, recognise that dredging and reclamation works require special consideration. However FIDIC have not given any consideration to dredging and reclamation work in the 1999 suite of FIDIC contracts, which comprise of:
- Red Book – Conditions of Contract for Construction for Building and Engineering Works Designed by the Employer,
- Yellow Book – Conditions of Contract for Plant and Design-Build,
- Silver Book – Conditions of Contract for EPC/Turnkey Projects,
- Green Book – Short Form of Contract.

FIDIC have traditionally produced contracts that endeavour to set out a fair balance on issues involving risk and payment. In the Red Book 3rd and 4th editions risks are generally borne by the party best able to control those risks. Payment is based on re-measurement of work actually performed and the contract is administered by an Engineer who must act impartially when exercising his discretion.

In the 1999 suite of contracts, FIDIC have focused on four distinct types of work scope, each with payment options ranging from lump sum to cost plus.

FIDIC do give guidance as to the suitability of each form of contract for various types of projects and it is clear that FIDIC do not intend the Conditions of Contract for Plant and Design-Build or the Conditions of Contract for EPC/Turnkey Projects to be suitable for dredging works.

As a reaction to the emphasis of the 1999 Red Book being on construction for building and engineering work, FIDIC in conjunction with IADC have prepared a Form of Contract for Dredging and Reclamation Works, which was published as a test edition in 2001. The implication of this is that it demonstrates FIDIC's acknowledgement of the special nature of dredging and reclamation works, which is not adequately addressed by the 1999 suite of contracts.

I understand that it is the intention of the FIDIC task group to publish the Form of Contract for Dredging and Reclamation Works 1st edition later in 2004. I hope that its use will assist in good project management of dredging works.

Contract Objective

It is essential not to loose sight of the fact that the contract is simply a record of the agreement between the parties. If it is a good agreement the contract will accurately record the understanding that the parties share as to their responsibilities, obligations and liabilities. This shared concept is not easy to achieve and many problem projects start with a contract that is signed by the parties, but each party has a different understanding of its and the other parties responsibilities, obligations and liabilities.

The basis of a construction contract is straightforward in that the Contractor executes and completes work in exchange for the payment of money by the Employer. However, when the consequences of change and non-performance are considered the contract becomes complex with many express terms.

Commercial success of the project will be promoted by contract documents that are clear and that fairly allocate risk.

FIDIC publish forms of contracts that are well known within the international construction industry. A former Chairman of the FIDIC Contracts Committee stated, in an introduction to the 1999 suite of FIDIC contracts, that FIDIC's "prime objective has been the preparation of contract documents the use of which will ensure completion of projects on time and within budget, while providing fair payment provisions for both employer and contractor"

The FIDIC Contracts Guide, published in 2000, identifies as a main feature of the 1999 Red Book that "the General Conditions allocate the risk between the parties on a fair and equitable basis: taking account of such matters as insurability, sound principles of project management, and each party's ability to foresee, and mitigate the effect of, the circumstances relevant to each risk."

These are good objectives if a project is to be well managed. Historically it has been a basic principle in FIDIC contracts that risks should be borne by the party best able to control those risks. This is a sound management principle as it is poor management to pass on a responsibility in the full knowledge that it will not be dealt with in an efficient and economic manner.

Elements of a construction contract

Construction contracts are required to deal with 4 fundamental elements:
- Scope of Work: for each project the work to be carried out is different.
- Location: the work is carried out at a unique location with local conditions not previously experienced.
- Time Scale: the time constraints vary.
- Price: the cost of the work is affected by the scope, location and time.

Each one of the above elements interact with the other elements and for each element it is necessary for any construction contract to include terms that establish the relevant obligations and liabilities of the parties to each other and to third parties.

The obligations and liabilities of the parties in respect of the elements of a construction contract are set out by the Conditions of Contract, which are in effect the rules that the parties agree to for the performance of the work.

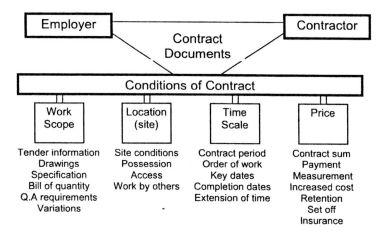

Figure 1. Elements of a construction contract

The conditions of contract that govern construction projects are often complex as a change in one element will impact on all other elements. For instance if the work scope increases then the site, time and price may also need to be increased. Good conditions of contract anticipate that changes may occur and set out express terms and procedures for the parties to follow.

Obligations / Liability / Benefits / Risk

The allocation of responsibilities for the elements and for changes to the elements impacts on the level of risk and reward that the parties have under a construction contract.

As demonstrated by figure 2 below, the levels of responsibility, risk and reward are primarily influenced by the work scope that the party is obliged to perform under the contract. When a contractor has a work scope incorporating financing to commissioning he will have a greater level of responsibility, risk and reward than if he has a work scope just for construction.

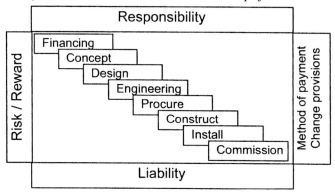

Figure 2. Work scope risk / reward

The extent to which the contractor's responsibility results in a risk or reward is determined by the conditions of contract and in particular the change provisions and the method of payment and measurement. A contract requiring the contractor to perform a broad work scope may be of low liability and risk if payment is calculated on a cost plus basis.

Types of Contract

Contracts are often described in terms of a combination of work scope and method of payment and measurement.

Work scope	Method of payment / measurement
• Construction	• Lump Sum
• Design and Construct	• Re-measurement
• Turnkey	• Cost Plus

There are many different published forms of contract, however very few forms of contract specifically take account of dredging and reclamation work.

FIDIC, the Federation Internationale Des Ingenieurs-Conseils, was founded in 1913 and has produced standard forms of contract for civil engineering works since 1957. The Red Book 3rd Edition (1977) and 4th Edition (1987) are conditions of contract for works of civil engineering construction that have and continue to be used as a basis for contracts involving dredging and reclamation.

Since the publication of the 4th edition there have been many changes in the construction industry particularly in the areas of financing and the attitude to risk management. In 1999 FIDIC published a new suite of four forms of contract that update previous editions and use a standardised style and format.

The 1999 suite demonstrate the types of contract commonly in use and for each type of contract FIDIC describe the scope and nature of the project that the respective contracts are suitable for.

1999 Red Book – Conditions of Contract for Construction for Building and Engineering Works Designed by the Employer
This is recommended by FIDIC for building or engineering works designed by the Employer or by his representative, the Engineer. Under the usual arrangements for this type of contract, the Contractor constructs the works in accordance with a design provided by the Employer. However, the works may include some elements of Contractor-designed works. The 1999 Red Book is now specifically aimed at "Building and Engineering Works" and unlike the Red Book 3rd and 4th editions there is no longer any reference to dredging works.

1999 Yellow Book – Conditions of Contract for Plant and Design-Build
This is recommended by FIDIC for the provision of electrical and/or mechanical plant, and for the design and execution of building or engineering works. As was the case for the Yellow Book 4th edition, the 1999 edition is not an appropriate form of contract for dredging and reclamation works.

1999 Silver Book – Conditions of Contract for EPC/Turnkey Projects
This is recommended by FIDIC for the provision on a turnkey basis of a process or power plant. It may be used where one entity takes total responsibility for the design and execution

of a privately financed infrastructure project, which involves little or no work underground. Under the usual arrangements for this type of contract, the entity carries out all the Engineering, Procurement and Construction ("EPC") providing a fully-equipped facility, ready for operation (at the "turn of a key"). This is an update of the Orange Book Conditions of Contract for Design-Build and Turnkey that was first published in 1995. I have seen the Orange Book used as a basis for a contract comprising dredging and reclamations work. The result was not a success and disputes continue. It is quite clear that FIDIC do not intend the Silver Book to be suitable for dredging works. Not only are dredging works wholly underground but they are also under water.

1999 Green Book – Short Form of Contract
Recommended by FIDIC for relatively simple or repetitive work, or work of short duration or of small capital value.

Special nature of Dredging and Reclamation Works

Red Book 3rd Edition – Part III
The Red Book 3rd edition (1977) comprises of 3 parts, Part I: General Conditions, Part II: Conditions of Particular Application and Part III: Conditions of Particular Application to Dredging and Reclamation Works. The following are some of the special considerations identified by FIDIC when the work comprises of dredging and reclamation:
- The Contractor is not normally held responsible for the maintenance of the Works after takeover.
- The Works are usually taken over in section as they are completed.
- The Contractor can only work economically if he is allowed to work continuously by day and by night.
- The incidence of Plant Costs (mobilisation, supply and demobilisation) forms a much higher proportion of total cost in the case of dredging contracts than is generally the case with construction contracts.
- As plant supplied by the Contractor almost invariably includes ships and at times includes ships taken on charter by the Contractor he cannot give to the Employer the unrestricted right to sell such plant.
- Quantities included in the tender documents must necessarily be estimates, the accuracy of which is inherently less than normally experienced on construction contracts.

Red Book 4th Edition – Part II
The 4th Edition did not retain the Part III from the 3rd Edition, however FIDIC state in the foreword of Part I that; "when dredging and certain types of reclamation works are involved special consideration must be given to Part II".

The introduction to Part II makes specific reference to dredging and reclamation works and states:
> "Special consideration must be given to Part II where dredging and certain types of reclamation work are involved. Dredgers are considerably more expensive than most items of Contractor's Equipment and the capital value of a dredger can often exceed the value of the Contract on which it is used. For this reason, it is in the interests of both the Employer and the Contractor that a dredger is operated intensively in the most economic fashion, subject to the quality of work and any other overriding factors. With this end view, it is customary to allow the Contractor to execute dredging work continuously by day and by night seven

days a week. Another difference from most civil engineering is that on dredging work the Contractor is not normally held responsible for the remedying of defects after the date of completion as certified under clause 48. Part II contains explanations and example wording to cover the above points and others relating to dredging. Clauses 11, 12, 18, 19, 28, 40, 45, 49, 50 and 51 are those which most often require attention in Part II when dredging works are involved and reference is included under each of these clauses. Other clauses may also need attention in Part II in certain circumstances. Reclamation work varies greatly in character and each instance must be considered before deciding whether it is appropriate to introduce in Part II changes similar to those adopted for dredging, or to use the standard civil engineering form unaltered"

FIDIC Form of Contract for Dredging and Reclamation Works

The absence of recognition of the special nature of the dredging industry in the 1999 suite of contracts was addressed by the publication in 2001 of a FIDIC Form of Contract for Dredging and Reclamation Works (test edition). It follows the format of the Short Form of Contract and was produced by FIDIC in collaboration with IADC.

In the foreword FIDIC state that its aim has been to produce a straightforward document which includes all essential commercial provisions, and which may be used for all types of dredging and reclamation work and ancillary construction with a variety of administrative arrangements. Under the usual arrangement for this contract, the Contractor constructs the Works in accordance with a design provided by the Employer or by his Engineer. However, this form may also be suitable for contracts that include, or wholly comprise, contractor-designed works.

Unlike the Red Book 3rd and 4th editions, that in their usual form are re-measurement contracts, under the D&R Contract the Employer has a choice of valuation methods.

The following is an overview of some of the conditions of the D&R Contract and how account has been taken of the special nature of dredging and reclamation works.

Agreement
The Agreement, which incorporates an Appendix, follows the format of the Short Form and is a single document that combines the offer and acceptance. FIDIC state that their aim is to promote a clear and unambiguous procedure of offer and acceptance that endeavours to avoid the traps and uncertainties that surround "letters of acceptance" and "letters of intent". In my experience many problem projects start with lack of clarity and misrepresentation at the time of contract. Any procedure that promotes clarity such that the parties to the contract have the same understanding of the obligations and liabilities being imposed is a welcome development.

The Employer – Permits, licences or approvals
It is a common trend in contracts to have provisions that appear to obligate a party to do something, but only to the extent stated elsewhere in the Contract. For example in respect of D&R Contract Clause 2.2, "the Employer shall obtain all permits, licences or approvals in respect of any planning, zoning or other similar permission required for the Works to proceed, as stated in the Appendix". If nothing is stated in the Appendix, the Employer does not have an obligation to obtain any permits, licences or approvals and the responsibility rests with the Contractor.

The Employer – Site Data
Reliable information relating to the expected conditions at the site are fundamental to the calculation of an accurate estimate. The D&R Contract does not quite follow in the footsteps of the 4th edition clause 11 in respect of the Employer obligation to make site data available to tenderers. A subtle change has been made in that the D&R Contract does not expressly include an obligation on the Employer to make available all information obtained on behalf of the Employer. There is a possibility that unscrupulous Employers will argue that information obtained by the Engineer and not in the Employers possession does not need to be made available to the Contractor. My view is that an Employer who knowingly withholds information that is relevant to the computation of contractor's tenders only has himself to blame when there are problems on the resultant project. I hope that when the 1st edition is published that this aspect will have been dealt with.

The Engineer
The D&R Contract retains a role for the Engineer and it is pleasing to see an express term that "the Engineer and any assistants shall exercise their duties and authority in a fair manner". I do not think that this goes so far as to mean that the Engineer's role is impartial as was the case in the 4th edition. It is interesting to note that the Engineer does not have a role in the adjudication of disputes.

The Contractor
The Contractor is responsible for the adequacy, stability and safety of all operations and of all methods of construction. If the Contractor is required to carry out any design it shall be fit for the intended purposes defined in the Contract.

Under the D&R Contract the Contractor does not have an express entitlement to work continuously by day and by night and on locally recognised holidays or days of rest. Working hours have a significant impact on cost and should be expressly agreed in any contract for dredging and reclamation works.

Defined Risks
D&R Contract Clause 6.1 gathers together in one place:
- the grounds for extension of time – Clause 7.3,
- the grounds for claims for costs – Clause 10.4 and
- the exclusions on the Employer's right for indemnity Clause 13.2.

The Defined Risks take account of the special nature of dredging and reclamation works and include amongst other provisions:
   h) Any operation of the forces of nature affecting the Site and/or the Works, which was unforeseeable or against which an experienced contractor could not reasonably have been expected to take precautions.
   i) Force Majeure.
   l) Physical obstructions or physical conditions encountered on the Site during the performance of the Works, which obstructions or conditions were not reasonably foreseeable by an experienced contractor and which the contractor immediately notified the Engineer.
   m) Climatic conditions more adverse than those specified in the Appendix.
   o) Damage, which is an unavoidable result of the Contractor's obligations to execute the Works.

Remedying Defects

Following the issue of a Taking Over Certificate there follows a period as specified in the Appendix during which the Contractor is responsible to remedy any defects. This is however subject to the terms of Clause 9.2 which states that "the Contractor shall have no obligation to remedy defects in the dredging works notified after the date on which the Works or Section were completed as stated in the Taking-Over Certificate but the Contractor's liability for the same shall remain unaffected". I understand this to mean that the Contractor has no obligation to retain his dredging equipment at the Site or return his dredging equipment to the Site in order to remedy defects, however the Contractor is still liable for any defects that occur.

Under the second paragraph of clause 9.1 the cost of remedying defects attributable to any cause, other than defects due to the Contractor's design, Materials, Plant or workmanship not being in accordance with the Contract, shall be valued as a Variation.

In some cases it is not commercially practicable to require the Contractor to maintain parts of dredging works that have been constructed to line and level as an early project activity throughout the whole contract period and therefore careful consideration of sectional completion is necessary in all projects involving dredging works.

Variations

It is good to see that the variation clause takes account of the special nature of the equipment used for dredging and reclamation works. The Contractor's position has been protected to prevent Variations being issued which could unfairly harm his trading position.

Variations cannot be ordered that:
- omit work in order to give it to another contractor,
- omit work in order to enable the Employer to carry out the work,
- require the mobilisation of further major dredging equipment unless the time and cost effect can be agreed in advance.

Claims

The Contractor is entitled to claim for costs as a result of any of the Defined Risks. The Employer may also claim if he considers himself entitled to any payment or deduction in connection with the Contract. This follows the trend of the 1999 Red Book that entitles the Engineer to take account of more favourable conditions when considering claims for unforeseeable physical conditions.

Contract Price and Payment

Unlike the Red Book 4th edition, which in its usual form is a re-measurement contract, under the D&R Contract the method by which the tender and contract price are to be calculated is optional and may be lump sum, re-measurement or cost plus. The payment and measurement options are reviewed later in this paper under the heading "Methods of Payment / Measurement".

Care of the Works

The Contractor is responsible for the Works prior to the date of the Taking-Over Certificate. He is protected if costs are incurred due to a Defined Risk and in respect of non-dredging work by the obligation to insure under Clause 14.

Limit of Contractor's Liability
A limit of the Contractor's overall liability is to be stated in the Appendix. FIDIC recommend that the limit should be the value of the Contract.

Insurance
The Contractor is required to effect insurance as stated in the Appendix. The Appendix excludes dredging work from the Works that the Contractor is required to insure.

Methods of Payment / Measurement
The notes for guidance to the D&R Contract explain FIDIC's view as to the circumstances under which the different forms of payment may be appropriate.
- Lump sum price:
  A lump sum offer without any supporting details. This may be suitable for very minor works where Variations are not anticipated and the Works will be completed in a short period requiring only one payment to the Contractor.
- Lump sum price with schedule of rates:
  A lump sum offer supported by schedules of rates prepared by the tenderer. This alternative may be suitable for a larger contract where Variations and stage payments are required and the Employer does not wish to prepare his own bill of quantities.
- Lump sum price with bill of quantities:
  A lump sum offer based on a bill of quantities prepared by the Employer. This is the same as above but the Employer prepares the bill of quantities, which the Contractor prices for the purpose of the valuation of any variations however the price for the original scope remains fixed.
- Re-measurement with bill of quantities:
  All tenderers price the items and estimated quantities detailed in a bill of quantities that is prepared by the Employer. The actual amount paid is valued based on the actual quantity of work performed and on the prices set out in the bill of quantities. In order to quantify the work performed a detailed method of measurement needs to be incorporated in the contract. This would suit a contract where the quantity of work is uncertain and many changes are envisaged to the Works after the Contract has been awarded.
- Cost plus:
  An estimate prepared by the tenderer which will be replaced by the actual cost of the Works calculated in accordance with terms set by the Employer. This would suit a project where the extent of work cannot be ascertained before the Contract is placed. In the tender documents, provision should be made for tenderers to indicate their allowances for overheads and profit.

The fundamental difference between lump sum and re-measurement contracts relates to the method of calculation of the price of the scope of work at the time that the contract is made. With a lump sum contract the price for the original scope of works is fixed, yet with a re-measurement contract the tender price is based on estimated quantities and is to be recalculated based on the actual quantity of work performed.

Traditionally dredging works have been contracted on a re-measurement basis due to the difficulty in predicting the nature and quantity of the material to be dredged. This however creates a need to have rules that determine how the work is to be measured. These rules are called the "method of measurement" and just like there are published standard forms of contract there are also many published standard forms of methods of measurement.

The aim of methods of measurement is to distinguish work with different cost considerations. For instance the nature of the material could have an impact on the cost as silt, sand and rock may all require different methods of working and may also have different settlement properties.

Generally the options are to measure quantities based on:
- Lines and levels shown on the drawings
- Actual lines and levels (in / out survey)
- Volumes in vessels
- Resource time expended (dayworks)

The choice of the method by which to calculate the quantities very much depends on a balance between accuracy, time and cost of measurement.

Conclusion

For any construction work successful completion for both parties will be promoted by:
- Contract documents that clearly record the agreement made.
- Contracts that allocate risk on a fair and equitable basis to the party best able to control it.
- Methods of payment that are fair, equitable and appropriate to the work scope.
- Methods of measurement that, given considerations as to cost and time, accurately quantify the work.

FIDIC continues to publish many forms of contract that they recommend for various situations and types of work.

FIDIC acknowledge the special nature of dredging and reclamation works and in 2001 published a Form of Contract for Dredging and Reclamation Works (test edition). The European Investment Bank has recently specified the use of this form of contract for a project in Albania.

As an industry we should endeavour to contract based on clear, equitable and fair conditions of contract and terms of payment. John Ruskin, a 19th century philosopher, eloquently puts the consequences of not doing so:

> "It is unwise to pay too much, but is worse to pay too little. When you pay too much you lose a little money. When you pay too little you sometimes lose everything, because the thing you bought was incapable of doing the thing it was bought to do."

> "The common law of business balance prohibits paying a little and getting a lot. It cannot be done."

Contract v in-house maintenance dredging

Paul Mitchell, UK Dredging, Cardiff, UK

Synopsis
Maintenance dredging operations can form a significant proportion of a ports total ongoing operating costs and in some cases it is the highest single item of port expenditure. Like most decisions made by any commercial organisation the decision whether or not maintenance dredging should be undertaken by equipment owned, operated, manned and maintained directly by the port (referred to as 'in-house dredging') or by engaging specialist dredging contractors to undertake the task (referred to as 'contract dredging') is primarily an economic one, however there are other factors that should also be taken into consideration.

Direct comparison between the benefits and disadvantages of in-house maintenance dredging and contract maintenance dredging to undertake what is essentially the same task is not always straightforward. The final decision can only be made after a thorough evaluation of all the factors involved and how these factors combine to produce the most economically advantageous solution for each situation.

Introduction
There are in the region of 105 commercial sea ports and harbours operating in the United Kingdom and Northern Ireland. Of these approximately 45 ports have a regular maintenance dredging requirement i.e. expect to undertake a maintenance dredging campaign at a frequency no greater than every three years. 18 ports utilise their own in-house dredging capability to satisfy the majority of their maintenance dredging requirement, that is dredging plant that is directly owned, operated, manned and maintained directly by the port operator. 23 ports routinely outsource their dredging requirement. This is usually achieved through the award of some form of contract made to a suitable dredging contractor who is generally selected through a competitive tendering process. 4 ports currently utilise a combination of their own in-house dredging plant that is supplemented when necessary by contractor's plant.

Because of the difficulties associated with the measurement and collation of data it is not easy to compile accurate figures that define the scale of the maintenance dredging commitment in UK ports and harbours in terms of the in-situ volume dredged. Various figures exist and one estimate, based on knowledge of the current UK maintenance dredging market, is that in the region of 20 million cubic metres in-situ per annum is dredged. Approximately 50% is undertaken by in-house plant and 50% completed under contract arrangements.

There has been much discussion regarding who is best placed to undertake maintenance dredging within ports and harbours and widely varied opinions exist depending on the past experiences and interests of individuals and organisations concerned. Typical comments that might be expressed on this subject would probably include *"in-house dredging plant is outdated and inefficient"* or alternatively *"dredging contractors will charge what ever they think they can get away with"*. If such comments are correct then this does not bode well for the port operator attempting to undertake maintenance dredging at the lowest cost with the least disruption to the commercial operations of the port. Perhaps historically such criticisms have been justified but fortunately, in reality today such views tend to be a legacy of past practices and miss understanding of the objectives of the respective organisations.

The long-term trend has been away from the use of in-house maintenance dredging and towards contracted out operations. The numbers of port owned dredgers has steadily declined. At the same time dredging contractors have evolved from small family-owned businesses financed by their own resources to major international concerns through organic growth, mergers and acquisitions.

Based on this trend it might be concluded that dredging contractors are more efficient than in-house dredging operators. This is not necessarily the case. The introduction of the Trailing Suction Hopper dredger or 'Trailer' initiated a fundamental change in the way maintenance dredging is undertaken. Trailers were first introduced into the UK during the late 1950's. Historically bucket dredgers and grab hopper dredgers undertook the majority of maintenance dredging. Because of their size, capacity, output and relative immobility each dredger was normally only able to service a single port or harbour, perhaps two or three ports if they were located in relatively close proximity and the dredging requirements were small or infrequent. Because of this dredgers were generally locally owned and operated usually directly by the port.

Trailers offered many advantages, being able to dredge and transport comparatively large volumes of material relatively quickly and efficiently while causing reduced obstruction to the port and being less susceptible to adverse sea conditions and hence the ability to work in more exposed locations. The sea going ability of trailers also provided a great increase in mobility of the plant at relatively low cost. With the obvious advantages of the trailer it soon became the standard tool for many maintenance dredging operations. International contractors quickly recognised that they could begin to challenge the dominance of locally operated dredgers and began to invest heavily in new plant that could be deployed worldwide. During the early 1960's the size and seagoing ability of trailers greatly increased and improved as did the numbers of trailers built [Ref. 1]. As a consequence of this developing trailer capacity by dredging contractors, the traditional port operated grab hopper and bucket ladder dredgers have largely been rendered redundant and in many cases they have been replaced by the more efficient contractors trailers. This is not always the case however and in the correct circumstances in-house operated trailing suction hopper dredges provide equally efficient dredging at the lowest cost to the port.

It is reasonable to propose that competition results in reduced costs and improved efficiency and that a dredging contractor operating in the open market must remain efficient in order to remain competitive and survive. However, it is also evident that competition is not the sole domain of the dredging contractor. Competition between dredging contractors is no more unique than the competition between ports [Ref 2]. This has necessitated that ports focus on minimising their operating costs and it is no longer viable for a port to finance an inefficient dredging operation. As a result the in-house dredging operation will be under similar pressures to those of the dredging contractor to continually strive to improve efficiency and provide the most cost effective dredging solution.

Unit Dredging Cost

The decision on how ports best undertake their maintenance dredging is not an easy one and a number of factors must be considered before arriving at the final solution. Ultimately the final decision will be based on an economic evaluation of the options available. There are various methods available for the measurement of dredging for payment purposes and cost comparisons. Discussion of the merits and disadvantages is outside the scope of this paper. The only satisfactory measure available to the port operator to allow direct cost comparison between in-house maintenance dredging and contract maintenance dredging is the total in-situ unit cost to the port i.e. the total cost per cubic metre of material dredged, transported and deposited measured in-situ at the dredge site. The in-situ unit cost will of course vary from port to port depending primarily on the site conditions and the plant employed.

Fundamentally, with knowledge of the production rates or output that a particular dredger can achieve at a given dredge site together with the total operating costs of the plant and its level of utilisation, it is relatively straight forward to calculate the in-situ unit cost of dredging. The in-house dredging operator needs to determine his plant operating costs, production and utilisation in order to establish an in-situ unit rate for dredging that can be costed against the alternative of using dredging contractors.

Given adequate information regarding the dredging task to be undertaken the dredging contractor will be in a position to provide his in-situ unit rate or lowest bid price based the most economic method of working using plant that is suitable to execute the work and is available from within the contractor's own fleet within the required time scale. The dredging contractors in-situ unit rate will be based on the contractors cost estimate. This is determined by the same three elements as the in-house unit dredging cost. Due regard will be given for mobilisation and demobilisation costs and depending on the market strategy of the contracting company the unit cost will be adjusted to account for profit, risk and perceived "contract value" before the unit price is submitted to the client. These adjustments may be positive or negative.

The level of profit that is included in the contractors unit price will depend on the state of the current market and the normal profit levels that can reasonably be expected to be achieved. The level of profit adjustment will also be influenced by a valuation of the likely competition taking into account factors such as plant suitability, availability, and location.

The evaluation of contract risk will largely be dependant on the type of contract being used and how the risks are distributed between the contractor and employer. Lump sum contracts generally imply that the contractor has to bear all the risks while charter contracts shift the majority of the risks to the employer. Contractual risks can be split into two groups. Technical risks include items such as delays due to weather, equipment failure, unexpected soil conditions and obstructions/debris. Institutional risks such as the risk default by the employer, exchange rates and inflation, political instability or strike action. The contractor will calculate his risks and include them in his pricing [Ref .3]. Some risks can be insured.

The dredging client will also incur additional costs as a result of contract dredging operations such as consultant's fees, supervision costs independent survey fees for measurement purposes and legal fees. All these costs should be identified and considered when determining the total unit dredging cost.

The Dredging Task

Regardless of who undertakes a ports maintenance dredging it is first important to adequately define the dredging task. The dredging task needs to be fully and accurately specified in order that sufficient information is available to be able to identify the most suitable dredging plant and determine the production rates that can be achieved. The dredging task will also influence the operating costs of the dredging plant since site conditions will affect factors such as fuel consumption and component wear rates and hence maintenance and repair costs.

The factors influencing the dredging task are different for each and every location and in the practical situation these factors are invariably interrelated. It is the interrelation of these factors that will determine the most suitable dredging plant. The following items should be considered when specifying the dredging task.

Geographical Location

The geographical location of the dredge area is important since it will impact on mobilisation and demobilisation time and cost of dredging plant. Availability of plant in the locality will result lower mobilisation and demobilisation costs and indeed for smaller dredging projects in the short term the availability of plant locally might outweigh plant efficiency for the specific task and result in the lowest bid from a dredging contractor. This is obviously of more consequence for contract dredging operations.

Geographical location will also determine the prevailing weather conditions that will impact on choice of most suitable plant and risk of weather related delays.

Frequency and Quantity

The quantity of material to be dredged can normally be specified in terms of in-situ volume to be dredged, the locations from where this volume is to be dredged and the frequency that dredging is required. This will depend on the natural regime of the site and the operational requirements of the port. Account should also be taken of any seasonal and annual variations in the volume of material required to be dredged.

Where little or no dredge tolerance is available and siltation is steady and ongoing then it may be necessary to undertake dredging on an almost continuous basis. Conversely, if larger tolerances can be accommodated or the natural regime results in high siltation on a seasonal basis then it may be possible that maintenance dredging takes place during a discrete campaign over a shorter period of time, leaving long periods when no dredging is necessary. In some circumstances the dredging requirement of a particular site may not be at all predictable in terms of frequency and quantity, for instance siltation may be entirely storm induced. In such circumstances it may be more difficult to establish a strategy that adequately reflects the requirements of the port.

Account should also be taken of the long-term development strategy of the port and how this will affect the future dredging commitment. For instance a planned capital dredge for a future channel deepening may result in an increased maintenance dredging requirement that is too great for existing dredging methods or prohibit its use due to depth restrictions and the like.

Soil Conditions

Maintenance dredging can encompass a wide variety of soils and a seemingly small change in soil description can influence the most economic method to execute the dredging. By definition of the term 'maintenance' the soil must be formed of particles that have been deposited since the original construction of the facility to be maintained, this can range from fine silts through sands to gravel, cobbles and even boulders [Ref. 4]. Soil conditions will affect the selection of the most appropriate dredging plant, the production rates that the selected plant is able to achieve and the wear rates that will be encountered. Soils conditions therefore need to be well understood in order to make any comparisons between the methods adopted to undertake a ports maintenance dredging.

Generally soil investigations for maintenance dredging operations are not given sufficient consideration and often consist of nothing more than a superficial examination and general description of the soil such as 'river mud'. Ideally as a minimum soil investigation should include the division of the different soil types over the area to be dredged together with the relevant physical properties of the soil. Further guidance is given in the PIANC publication Classification of Soils & Rocks to be Dredged [Ref. 5].

Ground Investigation may appear to be costly but should be considered against the total cost of the dredging. Furthermore the material encountered at a particular site is not likely to change significantly between one maintenance dredge campaign and the next and so one comprehensive soil investigation at the outset should be sufficient for future occasions.

The presence of debris, obstructions, chemical contamination and gas as a consequence of organic material should also be considered since this will have a significant effect on the proposed method and plant to be employed.

Bathymetry

Seabed levels and hence water depth available to the dredging plant to will influence the type and size of plant that can be utilised. Shallow water depths at the dredge site, deposit site or the transit between the two may limit the size of plant that can be used or restrict working during certain states of tide.

Working Environment

The working environments in which maintenance dredging is necessary can vary considerably. Even a single maintenance dredging project may involve the dredging of a relatively shallow, well sheltered area within an enclosed dock through to a harbour entrance and tight turning circle and long and exposed deep water approach channel subject to tidal variations, strong currents and heavy seas. These factors will influence selection of plant by restricting the size and type of plant that can be employed, the required manoeuvrability and the tolerance to sea conditions.

Port and Site Specific Requirements

The commercial aspects of the port cannot be overlooked when specifying the dredge task. For instance berth occupancy and commercial shipping movements within the port may cause delay to the dredging operation and the associated costs must be accounted for. There may be other factors such as obstructions and underwater pipelines and cables that can impact on the dredging.

Transportation and Placement

Transportation distances and placement options of the dredged material are integral parts of the dredging process and influence the type and size of dredging equipment employed. Generally longer sailing distances will favour larger plant. The placement options available will determine the type of discharge/unloading system required and may influence the type of plant available. This will be of particular importance if shore placement sites are to be used.

Environmental Aspects

Environmental considerations are justifiably playing an increasing role in all dredging projects and place constraints on how material can be dredged, transported and deposited. It is no longer acceptable to base the method of dredging solely on economic and engineering considerations. All parties involved in planning and arranging dredging operations should be aware of the local environmental issues and national and international legislation likely to effect the operation.

Environmental considerations most likely effect the choice of dredging plant employed include noise levels and water quality issues such as turbidity and dissolved oxygen content at both the dredging and disposal site. Environmental considerations may result in restricted working practices such as limiting time allowed for the overflowing of the hopper well of a trailer or completely change the way future maintenance dredging is undertaken for instance by prohibiting the dredging of large volumes of material from a site over a short period of time and taking the dredged material to sea and allowing the dredging of small volumes over longer periods of time and trickle feeding the material back into the estuary.

Dredge Plant and Production Rates

In an ideal world once the dredging task has been adequately specified the selection of the most suitable type and size of dredge plant can be made and the anticipated production rates determined. The general selection of plant available to both the dredging contractor and the in-house operator will broadly be the same. No one piece of dredging plant will be ideal for all areas within a port to be dredged and usually compromise will need to be made. In reality neither the in-house operator nor the dredging contractor will have an infinite selection of plant available to chose from.

Undoubtedly the experienced dredging contractor will have the benefit of working on many projects in various soil types and working conditions (including capital dredging projects). This coupled with ongoing research and development programmes will enable the dredging contractor to fine-tune the plant and methods employed and through innovation develop technology that continually improves the efficiency of the overall operation.

While the in-house operators dredge plant may be slightly less technologically advanced and sophisticated compared to the contractors plant, if fortunate enough to be able to build new dredge plant the in-house operator can take the opportunity to design, construct and equip a dredger that meets the specific needs of the port in terms of in terms of size (including hopper capacity, carrying capacity, dredging depth, length, beam and draught), power and manoeuvrability. The working parameters of the in-house plant will be known and the dredge plant can be tailored to suit.

This is also true of the equipment installed onboard the dredge plant. The dredge contractor will need to be prepared to work in all conditions and will therefore need to invest heavily in highly flexible plant that can be adapted to work efficiently in conditions yet unknown. If the in-house operator knows that gas has never been encountered within the area of operation or that dredged material will not be required to be pumped ashore then there is no need to invest in expensive degassing systems, bow discharge facilities or booster stations. In theory therefore there is no reason why the informed in-house operator should not be in a position to provide the most suitable dredge plant at a lower capital cost than the dredge contractor is able to provide.

Economies of scale are very evident in dredging plant. Generally increasing the size of dredging plant will increase the potential productivity of the plant but the capital investment and operating costs will increase at a slower rate. It follows that providing the larger plant can be effectively operated and adequately utilised then this should result in a reduced in-situ unit rate.

Ports already operating their own plant must regularly examine the efficiency of such plant and review it against the alternatives of using dredging contractors or investment in new plant.

The contractor's objective will not be to provide the ideal plant to execute the dredging task but rather to provide his lowest bid to the client based on the most economical method of working using plant that is suitable to execute the task and is available from within the contractor's own fleet within the required time scale.

Once the appropriate dredge plant has been selected the anticipated cycle times and production rates can be calculated.

Operating Costs

The dredging contractor will be well aware of his operating costs when preparing his cost estimate. It is important that the in-house operator is also fully aware of his true operating costs and that all costs are accounted for in order to allow true comparison between the various options available.

The operating costs of most dredge plant can classified under the following key headings:

    Manning costs (including employment costs, training and sickness cover)
    Fuel and Lubrications
    Consumable stores
    Maintenance and Repairs
    Supervision
    Depreciation
    Finance
    Insurance
    Overheads

There may be additional operating costs that will be incurred by dredging contractors that may be avoided by in-house operations. These include mobilisation and demobilisation, site set up costs and additional overheads to account for site visits, estimating, marketing and tendering.

The opportunity to share resource such as personnel function and finance with the port means that In-house dredging operations have the opportunity to reduce operation costs however care should be taken to ensure that the full costs of the resource allocated to the dredging operation is fully accounted for.

Plant Utilisation

Fundamental to the efficient deployment of any dredging plant is the ability to maintain adequate levels of utilisation. Investment in dredging plant is very capital intensive and it is very costly to have expensive plant, equipment and manpower stood idol.

Traditionally it is considered that dredging contractors have been able to maintain higher levels of utilisation than the typical in-house dredge operation. Historically restrictive local working practices and labour agreements often resulted in in-house dredging plant only being available to work for a maximum of 40 to 80 hours per week compared to a theoretical 168 hours per week for the dredging contractor. Such restrictions on in-house dredging plant have now largely disappeared and plant is able to operate on a similar basis to contractor's plant. Market forces in the labour market and national and international working hours legislation now dictate crewing arrangements.

There are of course other factors that affect the utilisation of dredging plant and both the in-house operator and dredge contractor must make an assessment of the utilisation that can be realistically achieved in order to determine the available productive time of the plant and the dredging unit rate.

Adequate allowance must be made for non productive time encountered as a result of mobilisation, demobilisation, maintenance, break down, crew change, taking fuel, water and provisions, adverse weather downtime and idle time due to lack of contracts. There is obviously a cost associated with this non-productive time that must be accounted for and included within the unit rate or identified as a separate cost e.g. mobilisation and demobilisation lump sums. It would normally be anticipated that in-house dredging plant is generally less susceptible to mobilisation, demobilisation and of course idle time due to lack of contracts.

All other things being equal, the key to a cost effective dredging operation is to determine the optimum type and size of dredging plant for a particular location based on the dredging task. It is then possible to assess if this dredging plant can be fully utilised while undertaking the maintenance dredging at that location. If this is the case then it is likely that the unit cost of dredging can be minimised by undertaking the dredging in-house. It may be possible to achieve sufficient utilisation by grouping of ports in joint ownership of dredging plant resulting in a shared resource.

Strategic Benefits

There may be other reasons for a port to maintain an in-house maintenance dredging capability that are of strategic benefit rather than pure economic rational.

A port's dredging may be considered so critical to its operation that the port may wish to take complete control and avoid the use of contractors completely. This will allow greater flexibility in the programming and timing of dredge operations allowing short-term prioritisation of dredging to suit commercial needs of the port that may not be possible under contract arrangements.

Plant reliability must also be a consideration. Major dredging contractors usually have an extensive fleet of dredging plant that can be called upon in the event of the designated plant becoming unserviceable. Faced with similar circumstances the in-house operator may have few alternatives and may need to call on the services of a dredge contractor.

Contractors rates are subject to the market forces of supply and demand – a port may decide that it does not wish to expose one of its major costs to the risk of such market fluctuations. The in-house solution can provide a more stable and predictable cost. Alternatively a port may decide it is not prepared to bear the technical risks associated with maintenance dredging and pass these directly to a dredging contractor.

Finance

The dredging industry is very capital-intensive and maintenance dredging is no exception. The capital costs of even a small sized modern Trailing Suction Hopper Dredger is likely to run in excess of £10 million. Tying up such a large amount of capital in just one piece of equipment is a very important business decision [Ref. 6]. If, after taking account of all the factors already discussed, it is demonstrated that cost savings can be made by a port employing its own in-house dredge plant, the implications of such a large capital investment must be fully understood and a capital appraisal undertaken. Such a large capital investment will reduce the ports potential to finance other commercial activities that may be of more direct economic benefit or support the long-term success and viability of the port. The opportunity costs of investing in dredge plant must be fully identified and considered. Before committing to such a capital investment the port operator must be able to demonstrate an internal rate of return in line with the required investment criteria of the port business.

Conclusion

The long-term trend has seen the numbers of port owned dredgers decline and a shift is observed away from in-house maintenance dredging plant and towards the use of dredging contractors. This is not necessarily because dredging contractors are more efficient at maintenance dredging than in-house operations. The growth in the dredging contractors share of the maintenance dredging market is explained by the developments that have taken place in the type, size and available technology of dredging plant used and in particular the dominance of the trailing suction hopper dredger in the maintenance dredging market. In many instances this has resulted in the dredger deemed most suitable for a particular port having a far great capacity than required. In such circumstances it may no longer be economic for a port to continue to operate its own in-house dredging plant.

Under the correct circumstances there is no reason why in-house dredging operation cannot be more cost effective than the alternative of using dredging contractors. Key to this is selection and provision of the most suitable plant together with a sufficient dredging commitment to maintain adequate levels of utilisation. Plant must be adequately maintained, operations sufficiently monitored and costs carefully identified and controlled.

Dredging methods, plant and technology is constantly developing and improving efficiencies. Contractors and in-house operators cannot afford to become complacent. Regular review is required to ensure that a port is adopting the most cost effective strategy to fulfil its maintenance dredging requirement and that capital in invested wisely.

References

1. STONE M — Can contractors be more cost effective than port plant
British Ports Association Seminar, March 1985

2. CHARLTON J C — The economics of port maintenance dredging by contract or direct labour
Proceedings of Institution of Civil Engineers Conference, Maintenance Dredging, May 1987

3. Training Institute For Dredging. Dredging Contracts and Costs

4. JACKSON P. J. — The selection of plant for maintenance dredging
Proceedings of Institution of Civil Engineers Conference, Maintenance Dredging, May 1987

5. PIANC — Classification of Soils & Rocks to be Dredged
Supplement to bulletin No 47, 1984

6. MORET W — Investment in the Dredging Industry
Terra et Aqua, March 1994

Modelling sediment transport and sedimentation

Dr M P Dearnaley, Estuaries and Dredging Group, HR Wallingford Ltd, UK
J V Baugh, Estuaries and Dredging Group, HR Wallingford Ltd, UK
Dr J R Spearman, Estuaries and Dredging Group, HR Wallingford Ltd, UK

Introduction

The need for maintenance dredging arises in areas where a required operational depth is reduced as a result of sediment accumulation from one or more sediment transport processes. Maintenance dredging may thus be an ongoing issue at an existing facility, a consideration for a future development or a temporary requirement occurring during construction works. Associated with maintenance dredging are a range of issues including operational, environmental and regulatory aspects. Other papers in this publication address these topics.

In many cases a sound understanding of the sediment transport processes occurring at a site and the ability to predict sediment transport under a changed circumstance, are required as a basis for addressing the issues linked to maintenance dredging activities. Examples include:
- Optimising the design of a facility to minimise sedimentation
- Prediction of siltation rates at a new facility to provide input to an environmental impact assessment of the development
- Prediction of the fate of fine material released into the water column during the dredging and disposal process as a means of managing these operations for efficiency or any potential environmental impacts

Sediment transport can take many different forms such as near bed transport, transport in suspension, littoral drift, fluid mud flow and channel side slope subsidence; these can all contribute to accumulation in an operational area and the subsequent need for maintenance dredging. The dredging process itself can also contribute to accumulation of material as a proportion of the material removed from the bed is released back into the water column and subsequently returns to the bed, potentially within the maintained area. The physical processes associated with the different sediment transport processes are subject to ongoing research and as a result are becoming increasingly well understood, although some processes more so than others. Useful gatherings of knowledge into sediment transport are contained within Whitehouse et al, (2000) for muds and Soulsby (1997) for sands.

Three techniques are available for assessing sedimentation and the requirements for maintenance dredging. These are field measurement, analytical or expert assessment and application of predictive modelling techniques. An integrated approach based on these three elements is usually adopted although there are circumstances where no new field measurements are required and situations where predictive modelling is unnecessary.

Ongoing research has resulted in a range of different models being developed utilising physical process understanding to provide a means of predicting sediment transport. The models vary widely both in the type of model (mud, sand, littoral drift etc.) and in the level of complexity of the model (in terms of the detail of the physical processes represented). This has lead to a situation where, except for the sediment transport specialists themselves, there is often perceived to be a bewildering range of transport models available for application to a particular issue. One of the roles of the sediment transport specialist is therefore to advise on the optimal approach to be adopted for a particular situation. The choice of approach must be scientifically based but clearly it will (and reasonably should) be influenced by any particular concerns identified, the degree of confidence required in the results, the finances available and the timescale for investigation.

This paper presents a tried and tested approach to assessment and prediction of sediment transport including the use of numerical models. In essence it provides a checklist for the specialist, the engineer, the Regulator and the client from which each can gain confidence in the results of the investigations. Two example studies, one on the Tamar Estuary and one on the Thames Estuary are used to illustrate elements of the outlined approach.

Approach

The approach summarised in Box 1 is derived from best practice and the long experience of the major international hydraulics laboratories who over the years have worked together progressing the understanding of the physical processes of sediment transport and developed and refined predictive tools to represent these processes. In the 1990's HR Wallingford, funded by the Department of the Environment Construction Directorate, produced a series of guidance documents relating to the prediction of sedimentation (Cooper and Dearnaley, 1996, Dearnaley et al 1997 and Odd, 1999). The outcome of the studies built upon best practice at the time and was basically an identification of the common steps undertaken and limitations that apply within any investigation of sedimentation at any location, be it a major container terminal or a small marina.

- Understand the client's need and the requirements of the Regulator(s)
- Review and analysis of available information
- Site specific identification of the dominant mechanisms for:
  - Sediment mobilisation
  - Sediment advection
  - Sediment accumulation
- Develop understanding of the system – formation of initial conceptual model
- Identification of techniques and tools to apply
- Identification of further data requirements
- Collection of further field data
- Set-up of appropriate models
- Calibration/validation of models
- Consideration of limitations and/or sensitivity of modelling
- Refinement of understanding of the system – improvement of the conceptual model
- Undertake predictive modelling ("what if?" tests)
- Synthesis of the results
- Presentation of the results to address client and Regulator requirements

Box 1 – Recommended approach to sediment transport model application

In addition to the development of the original guidelines the outcomes of the research have also influenced the general approach to prediction of morphological change within estuary systems (EMPHASYS Consortium, 2000).

Experience suggests that many of the items listed in the approach could occur in parallel and there is an amount of feedback between them, particularly as the understanding of the Regulators requirements may develop during the study.

Understand the client's needs and requirements of the Regulator(s)

The needs of the client will be apparent to the client, his engineer and other lead advisors working with the client. The requirements of the Regulator(s) and other interested parties whom the client chooses to involve will not be apparent without consultation. Thus there is a need to liaise with Regulators and other interested parties from an early stage (and throughout the investigations as required) to ensure that requirements of all parties are addressed. In some situations some of the requirements of the Regulators will only be clarified as investigations and understanding and solutions progress and become refined. For situations where an Environmental Impact Assessment is required the initial Environmental Scoping will usually be the basis for commencing the consultation. In the case of large new developments such as for the proposed London Gateway Port, this consultation was facilitated by Technical Group Meetings set up as part of the Public Inquiry process.

Review and analysis of available information

In some circumstances (for example extension of an existing major port facility) a vast amount of information is available concerning the hydraulic and sediment transport processes at a site of interest. In other situations (for example a green-field site in a developing country) there may be no reliable or relevant information available.

- Drawings and charts
- Information from geological and site investigation surveys
- Photographs and video of the site (including aerial and satellite images)
- Admiralty Tide tables and Pilot
- Bathymetric and topographic surveys of the site
- Bed sediment samples
- Data from local Harbour Authority on:
    - Harbour operations (lock operations, turning areas, ferry traffic, pilotage)
    - Past and present dredging (capital and maintenance)
    - Material types dredged
    - Locations and history of use of disposal sites
    - Sedimentation rates (annual, episodic and seasonal variations)
    - Sediment sources (fluvial, marine)
- Information held in Regulator's data bases
- Data from dredging Contractors who have worked at the site
- Anecdotal information from other users (e.g. fishermen, yachtsmen)
- Previous hydraulic studies (field measurement, analytical or modelling)
- Meteorological and oceanographic data (observational or forecast)
- Information from research programmes and publications (often sourced via the internet)

Box 2 – Information sources for sediment transport modelling

Sources of information are varied and the quantity or quality (reliability) of information is often highly variable. In some situations the client will be able to provide the bulk of the material from their own records. In other situations the information may have to be derived from other sources. Once obtained, further analysis of the information collated may be required.

Site specific identification of the dominant sediment transport processes

The purpose of collating available information is to aid the identification of the important sediment transport processes either occurring presently at the site or likely to occur following some proposed development. The different processes are listed in Box 3 under the headings of mobilisation, advection and sedimentation mechanism. The identification of the important mechanisms is required to develop the initial understanding of how the overall system behaves and whether further measurements should be made.

Mobilising Mechanisms	*Advection Mechanisms*	*Sedimentation Mechanisms*

• Wave breaking
• Wave stirring
• Fluidisation of the bed
• Fluid mud formation from settlement
• Erosion by currents
• Pick up by wind
• Fluvial input
• Re-suspension by dredging
• Re-suspension by vessel movements
• Side slope subsidence
• Bioturbation
	• Tidal currents
	• Fluvial flow
	• Near bed flow
	• Secondary currents
	• Density currents
	• Wave-driven flow
	• Littoral drift
	• Wind driven flow
	• Meteorologically induced flow
	• Vessel induced currents
	• Movement of fluid mud and other near bed high concentration suspensions
	• Mixing/dispersion of material in suspension
		• Reduction of:
		  - wave breaking
		  - wave-driven flow
		  - wave stirring
		  - tidal flows
		• Interception of:
		  - littoral drift
		  - fluid mud
		  - bed load
		  - wind load
		  - side slope subsidence
		• Deposition from suspension
		• Ecological stabilisation of sediments

Box 3 – Important sediment transport processes to be considered

Develop understanding of the system

The interpretation of the significance of the different mechanisms is based on a combination of scientific understanding and experience. The main issue in terms of assessing potential maintenance requirements is identifying what types of sediment are present in the vicinity of the site of interest and whether that sediment can be mobilised and advected such that it could accumulate at the site.

For example, if the bed material is clean sand and there are no sources of finer material (i.e. from wash load, cliff erosion or episodic fluvial inputs) then mud transport processes can be discounted. Conversely if there are no sources of sandy material on the bed in the vicinity of the site, sand infill can be ignored.

If the transport mechanisms have a seasonal or episodic nature then this needs to be considered in detail. Is sedimentation gradual over time, or is it associated with a short intense storm or fluvial event? If the former it could be considered manageable, if the latter consideration may need to be given to accommodating the ability to cope with episodic infill in the design and/or operation of the facility.

As the understanding of the system develops the aim is to arrive at an initial conceptual model including sediment sources, sediment sinks, transport mechanisms, etc. This initial conceptual model may, at least partly, include an element of hypothesis where aspects of the system are not clear from the available information.

Identification of techniques and tools to apply

Any technique applied to investigate sedimentation processes must be scientifically robust; i.e. it must be capable of incorporating the effects of the important sediment transport processes relevant to the particular investigation. For example sand transport physics should not be used in situations where fine cohesive material is the main sediment source.

The most cost-effective approach to an investigation of sedimentation at a particular location will nearly always be for an initial desk based assessment to be undertaken using the available data with appropriate assumptions made relating to data gaps. This desk assessment will inform the choice, if any, of numerical modelling tools to be used.

Following a desk assessment the study requirements may require the use of one or more numerical models. The choice of model should be based on the identified key processes, the level of complexity required to simulate them, the time and spatial scale of interest and the outputs required by the client and Regulators from the study. Box 4 below summarises the types of numerical model available for studies related to sediment transport.

Hydrodynamics	*Sediment transport*	*Bed change*
• 2D flow models • 3D flow models • Regional wave disturbance models • Local wave disturbance models	• 2D/3D mud models (in suspension) • 2D/3D sand models (bed load and in suspension) • Littoral drift models • Fluid mud models • Mixed sediment models • Sediment plume dispersion models	• 2D/3D morphological updating models (coupled back to hydrodynamics) • Channel profile models • Point models

Box 4 – Types of numerical model available for the study of sediment transport processes

The above list is not exhaustive, there may be instances were other techniques or tools may be used in studies related to sediment transport and maintenance dredging.

In the case of the Tamar the final choice of a 2D modelling study was a balance between the variable nature of the estuary (2D in Plymouth Sound and in the upper reaches, 3D in Hamoaze), the important transport processes and level of detail required for the study.

Further data requirements

From the collation of any available data and the initial conceptual model of the study area the next step within the approach presented is to assess the need for further data. The requirement for further data should come from the identification of gaps in the understanding of the key sediment transport processes within the conceptual model, the required accuracy of the assessment and the budget/time constraints for the study. In particular there are three common situations where further data may be required:

Where there is no existing data - Further data is most obviously required if no baseline information on the existing situation exists, although, if only an initial desk assessment is to be undertaken, experience of sites nearby may be sufficient.

To understanding key mechanisms - Key mechanisms identified or postulated by the conceptual model may require closer evaluation. For example, recorded rapid rates of infill with muddy material may suggest the presence of fluid mud, which should be further investigated at an early stage.

For calibration and validation of models - Any applied models used will need to be suitably robust in their representation of the exiting conditions and new data may be required to provide validation data. The definition of further data requirements should be informed by the choice of applied model and the time/spatial scale that will be simulated by the model.

Collection of further field data
The definition of further field data collection campaigns will depend on the level of knowledge regarding the key processes identified within the initial conceptual model, the time and spatial scale of the key processes and the overall budget/time scale of the study. Any new field data collection should keep in mind the purpose to which any collected data will be used and should complement existing data. The complexity of collecting the required data and its likely quality should also be taken into account, being aware of any errors inherent in the method used. For large or sensitive studies the collection of further data may be both comprehensive and expected. For such studies the collection of field data needs to be planned in advance at the first opportunity, particularly as the field experiments may need to take place at set times of the year, or indeed over the course of a year to obtain the relevant data.

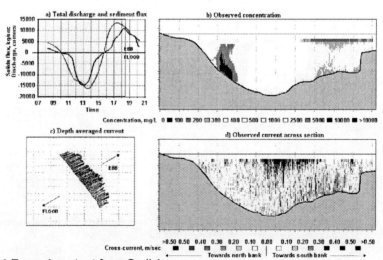

Figure 1 Example output from Sediview

One example of a new data type which was used to calibrate an estuarine sediment transport model was the Sediview method which was able to measure the total sediment flux through an estuary cross section (Figure 1). This ADCP backscatter based technique was applied to the Lower Thames Estuary and provided ideal calibration data for a 3D, coupled hydrodynamic and sediment transport model set up of the Lower Thames. The observations are fully described by DRL and HR Wallingford (2002). The observations enabled identification of cross-sectional variation in transverse currents and settling velocity as important influences in the 3D structure of sediment flux.

Set-up of appropriate models

With the choice of model to be applied to a sediment transport problem made and the required data either present or in the process of being collected the study focuses more closely on application of the model. Whatever the type of model the following considerations apply:

Model resolution

The discretization of the study area needs to be detailed enough to provide accurate representation of the important hydraulic features of the system such as the cross sectional area of an estuary and key features of interest to the client or Regulators (e.g. a dredged channel). However a balance has to be reached between accuracy and practicability, as ever-increasing resolution will result in overly cumbersome models.

Boundary conditions

The quality of data used to define boundary conditions requires consideration, as any errors resulting from poor boundary definition will propagate into the model domain. For example, tidal harmonics derived from a tidal record of 5 days is unlikely to provide data resulting in reliable tidal predictions within a tidal flow model. In many cases the data available for boundary conditions is either incomplete or non-simultaneous and then the establishment of complete, consistent boundary conditions can become part of the calibration process of the modelling.

In the case of the Tamar modelling studies, both the local effects of the works within the estuary and the fate of material disposed outside Plymouth Sound required study. Figure 2 shows the model grid used for the study.

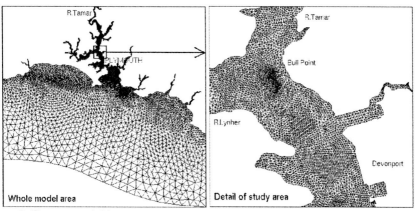

Figure 2. Tamar model layout

The model was used the unstructured grid, finite element model TELEMAC-2D which allowed detailed representation in the immediate area of the works at Bull Point, which keeping the imposed boundary conditions distant. The unstructured also enabled the offshore boundary to be aligned with the predominant directions of flow, simplifying definition of the boundary conditions.

Calibration/validation of models
Using available calibration data and appropriate calibration parameters (e.g. bed friction for flow models, erosion rate constant for sedimentation model, etc) a robust comparison between simulated and observed conditions should be produced.

The level of accuracy of the representation of the observation will depend of the nature of the observations (current, infill rate etc) and any data used should have appropriate levels of uncertainty identified. At the end of the calibration process one should, in general, expect good calibration and validation of flow modelling. For wave models direct comparison is not always possible but verification data for predicted wave conditions should be sought. Where data on sedimentation patterns and rates exists comparison of predictions and observations should be expected. Sedimentation data will often cover a longer period of time than that modelled with the possibility of conditions changing during the recorded period. However it should be possible to explain any of differences between model prediction and observation in terms of known changes to the system, e.g. changes to dredging practice, change in sediment supply, etc. The model may also be validated by simulation of historic conditions.

The model calibration is of course dependent on the processes included in it and any consistent problem in the calibration would lead to consideration of whether other mechanisms not included in the models might reasonably be expected to occur at the site.

In some cases objective statistical methods can help to identify the 'best' simulation when a model is compared to a large range of observations. The application of statistical methods to model calibration are described in Sutherland (2003).

Consideration of limitations and/or sensitivity of modelling
Bearing in mind the end use that the modelling outputs will be put to it is necessary to be able to attach confidence limits to the results. These should be carried forward into subsequent uses of the model results. Ideas of confidence come from understanding of the limitation of the modelling employed, the level of accuracy achieved in the calibration process and by the investigation of the sensitivity of the model results to changing important parameters be they physical (e.g. erosion constant) or numerical (e.g. choice of turbulence model).

Refinement of understanding of the system
The process of collecting further data, establishing and calibrating a model and the investigation of the limitations or sensitivities of a simulated system should feed back into the conceptual model. If the refined understanding of the system is considered to be robust (i.e. stands scrutiny of a range of specialists across a number of disciplines) it is then possible to proceed with predictive model runs.

Undertake predictive modelling (what if? tests)
In many cases sedimentation studies are required to predict conditions after a development or a change in dredging practice, under changed natural conditions such as changes in mean sea level. In these cases models are very effective, although the any identified limitations and sensitivity analysis above should be borne in mind in the presentation of predictive results.

Synthesis of the results
In general modelling studies produce very large amounts of information. This stage of the approach is aimed to reduce the information to that which informs the key concerns of the client and Regulators. For example a synthesis of a modelling study of the dispersion of a sediment plume from maintenance dredging may only be time histories of concentration at particular sensitive receivers or a plot of peak concentrations.

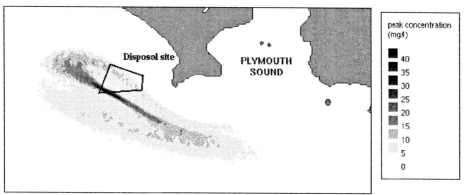

Figure 3 Peak concentration from disposal activities, spring tide

Figure 3 shows a typical synthesis of the sediment dispersion modelling results for the Tamar studies. While the model results provided full time histories of concentration at every model point it was the distribution of the peak, through tide concentration of suspended solids that was considered to be of most interest to the parties concerned with the disposal activities.

Presentation of the results to address client and Regulator requirements
Detail of the presentation such as minimum thresholds used for plots, even colours used require careful consideration. Plotted thresholds are a particular case for sediment dispersion results where the Regulators requirement for minimum increase in concentration should be considered. Along with the model results it is important to carry through the ideas of confidence and accuracy gathered at the stage of identification of model limitations and sensitivity analysis. For example, the accuracy of sedimentation predictions is unlikely to be greater than 30% to 300% except where good verification data on infill rates exists.

Conclusions
Appropriately designed and validated modelling studies can assess the magnitude and likely footprint of continuous, seasonal or episodic infill. They can also enable optimisation of management options such as sediment traps or changes to dredged areas and provide input to the dredging strategy e.g. over-dredging prior to monsoon season.

A variety of techniques exist to aid in the assessment of sedimentation in areas to be maintained. Confidence in the predictions from any technique comes from good quality data for the site/channel and good understanding of the sedimentation mechanisms.

The most cost-effective approach to an investigation of sedimentation at a particular location will nearly always be for an initial desk based assessment to be undertaken.
Models provide a useful, sometimes vital, tool for the assessment process although where possible evidence of model verification should be sought. The limitations, sensitivities and accuracy of any applied modelling technique should be presented along with the model results.

In the presentation of model results the aim should be to clearly address the issues that the client or Regulators wish to have informed by the modelling studies.

References

Cooper, A.J. and Dearnaley, M.P. (1996). Guidelines for the use of computational model in coastal and estuarial studies. HR Wallingford Report SR 456.

Dearnaley, M.P., Roberts, W., Spearman J.R., Cooper, A.J. (1997). Development of a unified approach for assessing sedimentation in Harbours. HR Wallingford Report SR 497.

Dredging Research Ltd and HR Wallingford (2002). The suspended solids regime at Coalhouse Point, River Thames, London.

EMPHASYS Consortium (2000). A guide to prediction of morphological change within estuarine systems. HR Wallingford Report TR 114.

Odd, N.V.M. (1999). Accuracy of marine siltation predictions. HR Wallingford Report SR 550.

Soulsby, R.L. (1997). Dynamics of Marine Sands: a manual for practical applications. Thomas Telford, London, ISBN 0-7277-2584-X.

Sutherland, J. and Soulsby, R.L. (2003). Use of model performance statistics in modelling coastal morphodynamics. Proceedings of the International Conference on Coastal Sediments 2003. CD-ROM Published by World Scientific Publishing Corp. and East Meets West Productions, Corpus Christi, Texas, USA. ISBN 981-238-422-7

Whitehouse, R.J.S., Soulsby, R.L., Roberts, W. and Mitchener, H.J. (2000). Dynamics of Estuarine Muds: a manual for practical applications. Thomas Telford, London, ISBN 0-7277-2864-4.

Predicting Environmental Impact

Peter Whitehead, ABP Marine Environmental Research Ltd, Southampton, England

Introduction

Dredging is highly regulated in the UK under a number of Acts of Parliament, predominantly by the Coast Protection Act 1949 (CPA) for both excavation and disposal operations and the Food and Environment Protection Act, 1985 (FEPA) for disposal operations in the marine environment. However, most Ports and Harbour Authorities have their own local powers to undertake maintenance dredging, particularly the excavation, but not always the disposal, which negates the need for CPA. This is not the case with respect to many of the marinas. The CPA invokes the need to consider the dredging with respect to the EU Council Directive 85/337/EEC (on the assessment of the effects of certain public and private projects on the environment) as amended by Council Directive 97/11/EC. This Directive is enacted with respect to dredging under the Harbour Works (Environment Impact Assessment) Regulations 1999. Both CPA and FEPA require consideration of the effects of the dredging with respect to Council Directives 79/409/EEC (Conservation of Wild Birds) and 92/43/EEC (Conservation of Natural Habitats and Wild Fauna and Flora). These are translated into British Law by the Conservation (Natural Habitats, &c.) Regulations 1994 (Habitats Regulations). Under this legislation, the UK Government interpret maintenance dredging, as well as capital dredging, as a project or plan and therefore requires each application for dredging and disposal to be assessed according to the strict procedure set out in the Regulations. In the future, maintenance dredging will also need to be considered with respect to the EU Water Framework Directive(2000/60/EC) and the forthcoming Environmental Liability Directive.

The general aims of such legislation are to:

- Protect the marine environment, the living resources which it supports and human health;
- Prevent interference with legitimate uses of the sea;
- Have regard to such other matters as the authorities consider relevant, such as noise, nuisance and odours;
- Protect/enhance nature conservation interests; and
- Generally ensure that maintenance dredging is undertaken, where possible, to the concept of sustainable development.

To comply with these legislative requirements, there is a need to predict the impact that all component parts of the maintenance dredging process have on the marine environment and its uses and users (both natural and anthropogenic). This paper sets out a framework aimed at identifying the likely areas of change and then assessing the significance of environmental impact, with respect to dredging activities (concentrating on maintenance activities). Details are given on the main potential effects of maintenance dredging and the impacts they can cause. Information on considerations and interactions which need to be integrated are also given to allow a reasoned argument for the severity of any impacts which may be identified. The conclusions draw out some of the 'over-arching' considerations which need to be taken into account when trying to predict the environmental impact from maintenance dredging.

Definition - Maintenance Dredging

The aims of the legislation are wide ranging, therefore to focus on the potential impacts of maintenance dredging, it is useful to consider its definition, which helps to define the areas and extents of potential changes and thus the receptors that might be affected. Maintenance dredging can be defined as:

> The routine periodic removal of any material normally deposited by the siltation process from port and harbour approach channels, basins, docks and marinas to maintain widths and depths to their design dimensions to ensure safe access for vessels.

In most cases, capital dredging will have been undertaken at some earlier time to create the original design dimensions. However, in some dynamic estuaries, 'high spots' may occur on the main navigation route, which require to be removed to ensure safe navigation outside of a specifically designated channel. Under FEPA, it is generally considered that, if an area has not required maintenance dredging within a period of 10 years, then the next dredge will be considered as a capital project. It should also be noted that the definition does not specifically mention the transport and disposal of the arisings. However, the term dredge is considered to cover these necessary activities.

Locations of Potential Change

Any consideration of potential impacts must cover the:

- Vicinity of excavation;
- Route corridor(s) between point of excavation and deposit/use location(s); and
- Vicinity of the deposit and/or alternative use within the marine environment.

At each location, effects of the dredging process will be both direct, e.g. removal of habitat at the point of excavation and potential sedimentation at the deposit location, or indirect, e.g. causing changes to the local hydrodynamics and sediment regime, which may lead to morphological change within the system. Depending on the relative locations of the individual areas, interactions between the individual effects may need consideration. Thus a major component in the prediction of environmental impact from maintenance dredging is identifying the extent of potential change around the site of excavation, the transport corridor, deposit/use area and the change to the local hydrodynamics in order to assess any indirect morphological effects.

Framework for Identification of Extent of Potential Change and Environmental Impact

Prediction of the extent of potential change and environmental impact from maintenance dredging requires information and assessment in three areas:

(1) An understanding of the dynamic nature of the hydrodynamic and sediment regimes within the marine environment, at least conceptually, in each of the potential locations

of change. This information provides a baseline against which changes introduced by maintenance dredging can be evaluated. Such an understanding can be significantly enhanced with the aid of a numerical model;

(2) Identification of the likely magnitude, frequency and duration of any change in the physical and chemical environmental parameters that will occur at each location, due to the dredging activity, both directly and indirectly. For example, volume and rate of dredge, and probable sediment release rates. This evaluation must take into account the material type along with the methodology and location of excavation, transport and disposal. At each location the dredging activity can cause a number of physical changes, which can lead to environmental impacts. These changes and resulting impacts are described in the following sections. Coupling of this information with the physical process baseline information from (1) above, either by desktop calculation or numerical modelling, will identify the extent of change, both spatially and with time, and determine the magnitude of change, for example in suspended sediment concentrations and/or sedimentation on the sea bed.

(3) Establish the baseline for the presence of individual receptors along with their natural variability and, if possible, any underlying trends that may already be occurring. The extent of information required will need to cover the area of predicted change determined from (2) above for each location. This information allows an assessment of the likely exposure for each receptor identified, which, when considered alongside the potential sensitivity of the receptor to that change, provides a prediction of impact.

In order to predict whether the changes caused by the maintenance dredging activities are likely to be an environmentally significant impact, it is necessary to integrate the above three sets of information to identify the areas of concern. This process should indicate the vulnerability of the individual receptors to the change. Often, due to a lack of information or gaps in knowledge, the consideration of whether the change is environmental significant is still a matter of expert judgment, where possible backed up by a 'risk based assessment', but always taking a 'weight of evidence' approach.

Dredging Effects in the Vicinity of the Excavation

Physical Changes
At the location of excavation, the main physical changes that will occur comprise:

- Removal of and disturbance to sediments;
- Noise and vibration in both air and water;
- Visual intrusion.

In some instances, the latter two may also impact on humans.

Sediment Disturbance
In most maintenance dredging, the greatest physical changes that may have potential environmental impact (in addition to the direct removal) are caused by the disturbance of sediments which can occur in a number of ways at the excavation location. At the sediment water interface, disturbance is caused by the action of the dredger cutting device, which can either be a mechanical action, as with a backhoe, grab or bucket dredger; predominantly hydraulic with a trailer suction hopper dredger (TSHD) or both, as is the case with a cutter

suction dredger. All machines are used for maintenance dredging but in the UK the majority of maintenance work is undertaken by TSHDs at the main ports and backhoe dredging in shallower areas, such as marinas.

Increasingly, more work is being carried out by 'non traditional methods', such as water injection dredging. Due to the different actions of the plant and rates or production, the amount of disturbance varies widely. For example, with water injection dredging, the aim is to move the material along the sea bed, therefore 100% disturbance occurs, whereas with some of the traditional methods disturbance rates at the bed can be below 5% of the amount of material dredged. The actual amount depends on the machine used, its size, cutter revolution rates, whether jetting is used and sediment characteristics.

Sediment is also released into the water column as the material is transported from the bed to surface when using the mechanical devices. This can occur due to wash off of sediment which adheres to the bucket/grab (both during rise and fall) or directly from the exposed surface of the sediment in the bucket.

At the surface, the sediment is normally loaded to barge or the hopper of the dredger. A considerable amount of sediment can be released into the water at this stage due to barge/hopper overflow in the process of improving the load, or through the use of automatic lean mixture overboard systems (ALMOBs). With older vessels these discharges were over the side of the vessel but newer craft tend to have an overflow weir, discharging through the vessel hull, i.e. below the water surface. The amount of material discharged from the vessel at the surface can vary considerably, depending on the operation of the dredging plant, to achieve the maximum economical production. Generally, reducing the overflow will reduce the dredger efficiency, however the economics is affected by the material type and distance to the deposit ground.

In some places, a cutter suction dredger is used where the material is pumped directly to the water column at a location usually in areas of high flow allowing the material to be naturally dispersed.

The above brief review of the dredging techniques show that, similar to the sea bed, rates of sediment disturbance/discharge at the surface can range from 100% of the material moved to the surface to almost no material, depending on the method employed. Bray et al (1997) gives extensive information on the areas of potential sediment resuspension from the different types of traditional dredging plant.

In addition to the actual act of excavation, further sediment disturbance can be induced by the manoeuvring of the dredging plant, particularly from propeller and bow thruster wash, and the effect of vessel blockage particularly in shallow confined areas where flow below the vessel keel can be considerable. Again, the amount of disturbance is difficult to predict as it depends again on plant type and size relative to the waterway being dredged.

Considering the number of ways in which sediment disturbance can be caused, it is almost impossible to monitor each one, although, with modern calibrated dredging monitoring instrumentation, fairly accurate calculations of the amount of sediment passed over the overflow of the TSHD can be obtained. Blockland (1988) devised a method of monitoring around different dredging plant to determine the total sediment release a short distance away from the excavation. From these observations, he was able to determine a ratio of the amount of sediment resuspended to the amount dredged, which was called the 'S' factor. This methodology can be used as a reasonable estimate of the amount of material which moves outside of about 50m of the dredge equipment (or the distance of the nearest reliable information). Kirby and Land (1991) reviewed published accounts of sediment monitoring around dredgers and derived 'S' factors for soft marine/harbour muds in a moderately active hydrodynamic environment. The data indicated release rates of the order of 3-30 kg of dry material for every m^3 dredged depending on the plant type and different modes of operation. It should, however, be noted that these rates are only indicative and actual rates depend on the sediment being dredged, the dredging method, local hydrodynamic and sediment regimes. The actual sediment disturbance rates are therefore very site specific and are likely to vary throughout the life of the dredge.

Noise and Vibration

The excavation of material causes both noise and vibration (or shock waves) within both air and water. Different dredging plant causes different types of noise. A grab or backhoe, for example, will cause engine noise from the vessel and equipment compressors above water level and impact noise from the bucket hitting the sea bed as well as the local shock wave in the water. This noise, however, will be intermittent depending on the bucket cycle time. Also, on smaller marina dredges, there may be long periods of inactivity whilst the barge deposits the material. A TSHD will produce vessel engine noise at the surface and noise of material moving through the pipe but little impact noise. Also, since the dredger moves to the deposit, the noise will be intermittent on the scale of hours, rather than minutes for a backhoe dredge.

Noise in the air has the potential to cause disturbance to local birds feeding or roosting, as well as to humans if the dredge is local to housing. This is most likely to be the case for mechanical dredging equipment, particularly bucket dredgers. However, very little the maintenance dredging in the UK is undertaken using such equipment.

Whilst all dredgers make some noise, this has to be considered against the ambient noise climate for the area. For example, Nedwell et al (2003) measured background noise levels in the vicinity of a piling operation in Southampton Water. The study indicated that background levels were 115-120 dB re 1μ Pa, which increased to around 140 dB re 1μ Pa due to dredger noise (TSHD) about 200m away.

Visual Intrusion

Visual intrusion may occur where dredging vessels are attendant for long periods of time. Again, the extent of visual change needs to be related to the baseline situation. The changes may not be as significant as sediment disturbance or noise, but in certain locations could be important predominantly with respect to the use of the area by birds.

Physical Change Along the Transport Corridor

Between the excavation and deposit location there can be physical changes due to:

- Leakage from the doors of the hopper or barge, or overspill in rougher weather. Generally, the amount of material deposited should be a very small proportion of the total dredged. The amount of leakage is likely to have an insignificant effect except if the leakage occurred over an area of very sensitive habitat;

- Secondary effects caused by the motion of the vessels, such as wash, particularly in narrow waterways. In most UK situations, this is not likely to be a major impact when removing material to a deposit ground. Transport to a beneficial use site, however, may require a route through an area where there is little other traffic, therefore vessel wash could assume some importance.

Physical Change at the Deposit Location

The deposit location for the sediment could be on land as well as the aquatic environment. More than est. 90% of all maintenance dredging in the UK is deposited within the marine environment, most at licensed disposal sites, although beneficial use is encouraged by the legislation. Generally, the effects are similar for both types of deposit below the high water mark and can be categorised as:

- Release of sediment in a large magnitude over a short period of time to the water column, sea bed/intertidal or both. Most maintenance dredging is deposited in a matter of a few minutes per cargo through bottom opening doors or from split hopper barges. Generally the material falls to the bed as a density flow, considerably faster than the component particle settling velocity, even for relatively low density cargoes. As it falls, a small proportion of material will be advected with the flow to settle out away from the deposit location. On impact with the sea bed, the local hydrodynamic process will try to disperse the material. At sites with strong hydrodynamics the material will be dispersed into the wider environment over a period of time. The effect is therefore the same as disturbance of material away from the excavation, except with much greater concentrations and shorter timescales (i.e. the order of tonnes of sediment as opposed to kilograms over minutes rather than hours). Most maintenance dredge material in the UK is fine grained, considerably less than 200 microns (median particle size) and therefore mobile in most environments. This can cause problems for beneficial use as generally the idea is to retain the sediment at the deposit location, particularly for intertidal habitat creation. The material is often deposited by pump out methods such as rainbowing from the vessel bow or returned to the sea bed down the pipe. These are generally a slower rate than bottom disposal but still orders of magnitude greater than natural processes. Dispersal of sediment to the wider environment from a beneficial use will most likely be less than for open water disposal;

- Disturbance of the sea bed due to direct impact from the deposited material. As the material impacts the bed the material may be re-suspended and moved away from the site adding to the volume of material moved. It is possible that higher density material dredged with mechanical plant, may stay in place and cause the surrounding bed material to scour away eventually burying the deposited material. This effect is more likely from capital dredged material rather than maintenance. However, if it does occur, it could cause a significant change in the material type of the bed.

- Run off scour. When material is placed on the intertidal, it is generally in a form containing in excess of 70% water. This water then runs off (or is weired off if the material is contained). It is possible that such run off could cause scour of the lower areas of the existing intertidal, thus potentially negating the effect of the beneficial use. Such effects need to be foreseen and managed to ensure that the perceived benefit is achieved without causing further changes at other locations.

Potential Morphological and Sediment Budget Change

The maintenance dredging will also cause some volumetric change to the system and temporarily change the sediment budget, although it is a matter of debate about whether such impacts should be ascribed to the maintenance dredge or to the original capital dredge. The dredging has the potential to change the local hydrodynamics and therefore may affect the local erosion and accretion patterns, creating the potential for habitat impacts, i.e. direct loss of intertidal or subtidal habitats, or smothering of the bed away from the area of the dredge activity. The amount of change will depend on the size of the dredge relative to the size of the cross-section and the estuary as a whole. A relatively small dredge in a small estuary may have a greater effect than a considerably larger dredge in a large estuary. Whether any hydrodynamic change causes a significant impact will depend on how close the flow regime is to thresholds of erosion and deposition of the local material types. From a morphological perspective, it is better to undertake the maintenance dredging on a little and often basis rather than a larger amount less frequently. In the former case, the change due to the dredge is smaller, creating smaller perturbations in the system. However, trying to predict the effect of these small changes is very difficult, even with the aid of numerical models, as the results are often of the same size as the inherent errors in the modelling process and therefore identifying the magnitude of change is difficult. The modelling of larger changes will give a greater certainty of the results. This becomes very important when trying to ascertain the potential indirect impacts of maintenance dredging on local intertidal areas or with respect to nature conservation designations.

Maintenance dredging may also remove sediment from the system, thus changing the sediment budget. This occurs when the sediment is removed to an offshore deposit site. However, where the sediment is simply relocated within the system, in the long term, this may assist in maintaining the sediment budget. Whether these effects have significance will depend on whether the system actually needs that sediment to develop, which is often difficult to establish in the short-term, particularly in highly dynamic environments which are subject to sea level rise.

Environmental Impact

The most significant environmental impacts caused by maintenance dredging occur to the marine habitats in the areas influenced, as described above. The following sections first briefly describe the main potential habitat impacts that could occur due to the resulting changes of the maintenance dredging process. Whether these impacts actually occur depend on the distribution and importance of the various receptors, identified in part 3 of the framework for impact prediction, given earlier. Even if habitat impact occurs, there is a need to determine the significance of such impacts. The considerations and interactions for this assessment are described in a later section. Other impacts may be associated with changes in noise or to visual intrusion.

Potential Habitat Impacts

Maintenance dredging causes direct habitat impacts, the main ones being:

- The removal of the bed which can change the characteristics of the bottom sediments and therefore the fauna which can live in the area;

- Direct removal or destruction of the benthos. This is only likely to be a problem where the maintenance dredging requirement is small and therefore is only repeated after a number of years. This time period allows the potential for plants and animals to develop in the accumulating sediments. In areas where dredging occurs a number of times a year, the development of plants and animals in the area is less likely;

- Possible relocation of fauna from the excavation to the deposit location. This, however, is likely to be limited to dredging undertaken by mechanical techniques. It is unlikely the benthos will survive through a dredge pump.

As identified earlier, the main effects of maintenance dredging are from all forms of sediment disturbance and this is the 'root cause' of most concern within the marine environment. Sediment disturbance initially causes a potential change to the water column quality by:

- Increasing the turbidity, i.e. reducing the penetration of light;
- Increasing the total suspended solids (both organic and inorganic);
- Increasing the bio-availability of any contaminants in the sediment, due to the release of adsorbed heavy metals or other toxic substances, therefore increasing the potential for impacts to living organisms.

The increase in turbidity and suspended solids in the water column have the potential to cause detrimental environmental impacts by:

- Reducing dissolved oxygen levels (deoxygenation) that can lead to suffocation of plants and animals. Benthic communities and fish require oxygen in the water to live, however, they all have differing tolerances to change. A sudden drop in dissolved oxygen may cause mortality but slower rates may lead to avoidance mechanisms. Mobile animals will probably move to another area but may return when oxygen levels increase. Sedentary species can withstand changes for short periods but, if effects persist, toxic effects are likely to result. Such changes can have 'knock-on' effects since a food resource for birds may be depleted or moved to another area. In the UK the Environment Agency (Stiff et al, 1992) have set Environment Quality Standards (EQSs) for dissolved oxygen (DO) for sensitive aquatic life e.g. salmonids or fish nursery grounds. This states that DO concentrations should exceed 5 mg/l for 95% of the time and 9 mg/l for over 50% of the time.

- Reducing photosynthesis by phytoplankton, phytobenthos or macro algae, which in turn can have an effect on the health of some benthic communities. The amount of turbidity is a function of the type and intensity of the dredging operation combined with the characteristics of the material type and the local flow conditions. Whether it is a significant impact will depend on the natural background turbidity and the presence of the flora and fauna likely to be affected;

- Increasing the bio-availability of contaminants, leading to increased bioaccumulation in the water column living organisms;

- Causing interference with feeding and respiration of benthic fauna and the blocking of gills in fish.

The effects of maintenance dredging induced sediment disturbance also impact on the sea bed. Most disturbed sediments resettle covering the sea bed around the extraction and/or deposit sites. This can lead to:

- Possible reduction/elimination of food supply, e.g. for fish (and/or birds over intertidal areas);
- Reduced habitat diversity;
- Increased chemical contamination of the area, should the sediment contain such contaminants. This is controlled by the legislative process, however, no sediment is completely clean and accumulations in some areas can occur. There are at present no Environmental Quality Standards for contaminants in sediments in the UK, however there are with respect to the water column. These can therefore act as a control on the maintenance dredging process should disturbance of sediments containing some contaminants occur.

Whether a significant impact of smothering of bottom dwelling organisms will occur depends on the rate of settlement, the overall thickness and the material type. Maurer et al. (1981a & 1981b) carried out experiments on the lethality of sediment overburden on selected macro-invertebrates. They concluded that many mobile epibenthic and infaunal animals could withstand a light overburden of sediment, especially when the overlying sediment was native to their habitat. They also found that increased depth and frequency of burial increased the rate of mortality. In addition mortality was also linked to water temperature, being greater during summer than winter. These findings are supported by the work of the Oresund Konsortiet (1998) who found that the infaunal bivalve communities could tolerate sedimentation rates of the order of 2 – 7 cm per month.

All these impacts could potentially occur around the sites of excavation or disposal/use but the severity of any impact depends on the magnitude, frequency and duration of the change in conjunction with the presence of receptor species and variability within the hydrodynamic regime.

Numerical modelling of the disturbance of sediment away from the site of excavation or disposal is a valuable tool in determining the extent of any effect from the individual activity. This will also define the areas where consideration of the various receptors can be focused. The modelling can also indicate the likely levels of concentrations of sediment in the water column and the rate and location of sediment build up on the bed from the dredging process. Such modelling is therefore a valuable tool in determining the order of magnitude of environmental impact. Care is however needed in interpretation of the results taking into account the uncertainties in the many input parameters, the simplification of the simulation and modelling errors. Often it will be necessary to undertake a series of sensitivity runs.

Noise impacts
In water, the main impacts of noise occur to fish and marine mammals, however as with other impacts the relative change against the ambient conditions as well as the species of concern need careful consideration. Some of the impacts of increased noise include disorientation, damage to swim bladders and eyes and interruption of their passage around the dredging plant.

Turnpenny and Nedwell (1994) carried out a review of the effects of marine fish, diving mammals and birds due to underwater sound generated by seismic surveys. This study indicated that serious injuries to fish (eggs to adults) only appear to occur at sound levels of the order of 220 dB re 1µ Pa but fish avoidance typically occurred at levels of 160-180 dB re 1µ Pa. Nedwell et al (2003) measured underwater noise during a piling operation in Southampton Water and the reaction and signs of injury in caged trout at locations away from the source. The impact piling created noise at a level of 194 dB re 1µ Pa one metre from the source, which had a transmission loss rate of about 0.15 dB per metre. 400m from the source (134 dB re 1µ Pa) there was no evidence of injury or reaction to the piling. Using the transmission loss rate this would have given a noise level of about 170 dB re 1µ Pa one metre from the source, i.e. about 50 dB above the background. Data from the previous report would indicate fish would just seek to avoid the dredger but would not be damaged. It should also be noted that the dredger did not cause greater noise than other commercial vessels. Moreover, conditions will change from place to place, particularly with respect to the transmission loss rate. Such rates derived at one location should only be seen as indicative and it does not mean that impacts recorded (or not recorded) at one site will necessarily be the same at another. Also, the animal receptors may be completely different and 'attuned' to different ambient conditions.

Thus, the same dredging plant at different locations will have a different impact dependent on the use of the local area above water and the species of animal living within the water.

Visual intrusion impact
For most maintenance dredging the impact of visual intrusion may well be negligible. At certain locations, for example where dredging or disposal takes place in close proximity to a mudflat or saltmarsh it may become a consideration that needs to be evaluated. This is likely to be the case for a static dredger with attendant barge for long periods (a number of hours) which could cause stress to birds feeding and roosting due to the interruption of lines of sight thus forcing birds to move to another location. Such effects are only likely to be temporary but may be considered significant at certain times of the year. The level of importance will vary again from place to place depending on the individual receptors in a particular area.

Evaluation of significance of impacts
All maintenance dredging has some effect on the aquatic environment, mainly due to an impact on the habitats and species in the affected areas. Whether these impacts are significantly detrimental (or beneficial) to the habitats and species depends on the interaction of many factors, including:

- Natural variability of the local hydrodynamic, sediment and morphological regimes. This can vary widely at different locations within and between estuaries. For example, the Humber Estuary is very dynamic with substantial subtidal changes in large areas occurring over periods of days on the back of morphological changes on the timescale of many years. This leads to a large range in suspended sediment concentrations both in time and space. On the other hand, Southampton Water is by comparison a stable estuary with a relatively small range in natural background sediment concentrations. In fact in the UK, it is unlikely that any two estuaries can be considered alike. In determining the significance of any impact caused by maintenance dredging the scale of the change must be considered against the local natural variability within the system;
- Physical and chemical characteristics of the dredged material relative to the same parameters in the receiving environment. If these are different, the deposit of the dredged material (either directly or due to sedimentation) could change the character of the sea bed and therefore animals and plants living in the area, i.e. causing ecological change. Again the significance will depend on the natural variability and the magnitude of the event;

- Type of dredging equipment used and the methodology employed. This choice depends on the scale of the dredge, type of material, and location of the dredge and disposal. The equipment and its use will also determine the rates of sediment disturbance at the excavation site and rate of discharge, character of material and frequency of discharge at the disposal location. All these parameters influence the severity of any habitat impacts at the various locations affected;

- Timing of the dredging activity. The relative impact on different receptors will vary throughout the year, therefore it is important to identify the receptors at risk in the locations that will be affected by the dredge. On shorter timescales, extents of effect can be reduced, for example by avoiding dredging or disposal at times of maximum flows, or restricting activity to flood or ebb tides. Restriction of dredging to certain times is one method used as an impact reduction measure. However, its use is often based on the perception of an impact rather than a true evaluation of the real level of concern;

- Plant and animal communities within the zone of influence. The level of impact will depend on the type, uniqueness, health and abundance, as well as any conservation designation status combined with the natural variability of their habitat. Consideration of these parameters will indicate the tolerance of the plants and animals to change. This allows an evaluation of the likelihood of recovery from the perturbation caused by the dredging activity;

- With respect to indirect effects, such as changes to the hydrodynamics, the impact will be in part determined by the existing local morphology and geology, particularly if increases in flow are induced in a particular area.

The considerations required in evaluating whether the effects of maintenance dredging will cause significant environmental impact are shown above to be extremely complex. It is very unlikely that all the information will be available to make a judgement and therefore a level of uncertainty with any evaluation will always exist.

Conclusions

In the UK, dredging is highly regulated through various legislation, however, greater focus on environmental impacts, particularly with respect to habitats, has occurred since the introduction of the EU Habitats and Birds Directives in the early 1990s. This legislation has been interpreted in the UK to require an assessment of the likely significant effect of the maintenance dredging activity on each licence application.

Predicting environmental impact from maintenance dredging is, however, a very complex topic where impacts can occur around three discrete locations for any activity; the site of excavation, the transport corridor and the site of deposit/use. Depending on the relative locations to each other, any effects may interact. To be able to identify extents of potential change or effect and then determine whether these result in a significant environmental impact, there is a need to:

- Understand the dynamics of the system;
- Identify the extent, magnitude, frequency and duration of the effects of all phases of the dredging process;

- Determine the receptors of concern which are likely to be exposed to the dredging effects and their sensitivity to change;

and then integrate the results.

Dredging cannot be undertaken without some effect on the aquatic environment, however, it is unlikely that all information will be available for the evaluation of environmental significance without some uncertainty. Therefore, a combination of expert judgement, risk assessment and a weight of evidence approach will be required. In any evaluation, the complexity and sheer number affecting parameters requires a site and event specific approach. Therefore, conclusions drawn from one location or dredge activity should not be used for another.

Most environmental impacts from maintenance dredging have the 'root cause' of initial disturbance of sediment to the water column, either from the dredging or indirect change to the natural system processes. However, whether these effects cause significant environmental impact depend on the relative scale compared to the natural variation of the characteristic hydrodynamic, sedimentological and biological parameters for the site. Minimisation of impact in a practical manner is therefore a function of trying to minimise the need for dredging, then selecting the equipment appropriate to the location that can be operated to minimise sediment disturbance, as well as having an understanding of the areas that are likely to be impacted and the existing habitat importance.

Due to the complexity of the possible interactions, any standards, mitigation and monitoring measures developed for one location will not be appropriate, necessarily effective or even needed at another location. The prediction of environmental impacts from maintenance dredging is site and event specific.

References

Blockland, T. (1988). Determination of dredging induced turbidity. Terra et Aqua, No 38, pp 3-12.

Bray, R.N., Bates, A.D. and Land, J.M. (1997). Dredging. A Handbook for Engineers, Second Edition, Arnold, London.

CEDA/IADC (1996-2001). Environmental Aspects of Dredging, Series of Guides:
1. Players, Processes and Perspectives.
2. Conventions, Codes and Conditions; Marine and Land Disposal.
3. Investigation, Interpretation and Impact.
4. Machines, Methods and Mitigation.
5. Reuse, Recycle or Relocate.
6. Effects, Ecology and Economy.
7. Frameworks, Philosophies and the Future.

Kirby, R. and Land, J.M. (1991). The impact of dredging - A comparison of natural and man-made disturbance to cohesive sedimentary regimes. CEDA - PIANC Conference - Proceedings of Accessible Harbours. Session B, Paper B3.

Maurer, D., Keck, R.T., Tinsman, J.C. and Leatham, W.A. (1981a). Vertical migration and mortality of benthos in dredged material – part I: Mollusca, Marine Environ. Research, 4: 299-319.

Maurer, D., Keck, R.T., Tinsman, J.C. and Leatham, W.A. (1981b). Vertical migration and mortality of benthos in dredged material – part II: Crustacea, Marine Environ. Research, 5: 301-317.

Nedwell, J.R., Turnpenny, A.W.H., Langworthy, J. and Edwards, B. (2003). Measurements of underwater noise during piling at the Red Funnel Terminal, Southampton, and observations of its effects on caged fish. Report from Subacoustech Ltd to Red Funnel, Reference 558R0207.

Oresund Konsortiet. (1998). The Oresund Link, Assessment of the impacts on the environment of the Oresund Link.

Stiff, M.J., Cartwright, N.G. and Crane, R.I. (1992). Environmental Quality Standards for Dissolved Oxygen. WRc and NRA R & D Note 130.

Turnpenny, A.W.H. and Nedwell, J.R. (1994). The effects on Marine Fish, Diving Mammals and Birds of underwater sound generated by seismic surveys. Report by Fawley Aquatic Research Laboratories Ltd, FCR 089/94.

Legislation

HMSO (1985)	Food and Environment Protection Act, 1985.
HMSO (1999)	The Harbour Works (Environmental Impact Assessment) Regulations 1999, Statutory Instrument 1999 No 3445.
HMSO (1949)	Coast Protection Act, 1949.
HMSO (1994)	Wildlife Countryside - The Conservation (Natural Habitats, &c.) Regulations, 1994, Statutory Instrument 1994 No 2716.
EU (1985)	Council Directive 85/337/EEC (on the assessment of effects of certain public and private projects on the environment) as amended by Council Directive 97/11/EC.
EU (1979)	Council Directive 79/409/EEC (Conservation of Wild Birds).
EU (1992)	Council Directive 92/43/EEC (Conservation of Natural Habitats and Wild Fauna and Flora).
EU	Council Directive 2000/60/EC (Water Framework Directive).
EU	White Paper on Environmental Liability (COM(2000)66 Final).

Planning for Environmental Protection

Nicola Clay, Environmental Scientist, Port of London Authority, United Kingdom.

Introduction
Maintenance dredging is an essential feature of the operation of many ports, harbours and marinas. By its nature maintenance dredging is an ongoing activity with site-specific characteristics, for example, the type of dredging methodology and its frequency (ranging from daily or weekly operations through quarterly to annual or greater periods). Each dredging location will also have individual environmental sensitivities and constraints which will need to be considered when a maintenance dredge is planned.

This paper draws upon experience gained in the Thames Estuary to present an account of the key considerations for building environmental protection into maintenance dredging operations. These considerations are applied on the Thames Estuary by the Port of London Authority (PLA) as part of its assessment of applications for maintenance dredging licences. The importance of building positive relationships with stakeholders is explained and the need for sharing of understanding of environmental sensitivities and economic constraints is highlighted.

Planning maintenance dredging to meet environmental best practice is a complex process, often involving understanding and compromise on behalf of the operator and environmental stakeholders. This paper aims to provide an insight into both the environmental and economic aspects and to offer a Regulator's view on achieving an appropriate balance as a contribution towards sustainable development.

Regulation of maintenance Dredging
Maintenance dredging is regulated in England and Wales by two mechanisms: consent under the Coast Protection Act 1949 and/or consent under a local Harbour Act. Environmental protection relating to the effects of maintenance dredging is provided by the general environment requirements on Harbour Authorities as set out in section 48a of the Harbours Act 1964, the Harbour Works (EIA) Regulations 1999 and the Conservation Regulations 1994. Further, environmental legislation relating to specific issues or features must also be considered, for example, the various Dangerous Substances Regulations.

Role of the Port of London Authority
Dredging in the Thames Estuary, within the limits of the Port of London (Teddington Lock to a line between Clacton and Margate) is regulated by the PLA by means of an application under Section 73 of the Port of London Act 1968. When considering a dredging licence application, the PLA is required by Section 48a of the Harbours Act 1964 to '*have regard to:*

a) *the conservation of the natural beauty of the countryside and of flora, fauna and geological or physiographical features of interest;*
b) *the desirability of preserving for the public any freedom of access to places of natural beauty; and*
c) *the desirability of maintaining the availability to the public of any facility for visiting or inspecting any building, site or object of archaeological, architectural or historic interest.'*

Consideration of the environmental sensitivities in an area and the applicable environmental legislation should be the first step in both planning and assessing a maintenance dredging operation. The environmental sensitivities will usually include, *inter alia*, designated conservation sites, sediment quality, water quality, fisheries (spawning and juveniles), marine biology, birds and archaeology. In the context of the type of dredging technique, the potential effects on hydrodynamics and sediment transport will also be key in considering effects on the environmental features.

Each of these issues will need to be considered during the PLA's application assessment process. However, the PLA will also give consideration to the cost-effectiveness and commercial affordability of any environmental requirements. The Authority aims to strike a balance between environmental protection and a thriving port, in accordance with the theme of sustainable development.

Maintenance Dredging Framework

In recognition of the need to demonstrate an open, transparent and inclusive decision-making process, the PLA, working with the Dredging Liaison Group of the Thames Estuary Partnership (TEP), has developed a Maintenance Dredging Framework. The Maintenance Dredging Framework comprises a number of different components as follows:

- An Information Exchange System;
- Environmental impact assessment and appraisal procedures;
- Beneficial uses register;
- Information notes for berth owners and operators;
- Consultation mechanisms;
- Data collection and monitoring; and
- Collaborative research.

Dredging Liaison Group

Discussion and consultation about maintenance dredging on the Thames is facilitated by the TEP's Dredging Liaison Group. This group comprises regulators, statutory advisors, environmental stakeholders and industry representatives. Dredging Liaison Group members are listed below:

- Central Dredging Association (CEDA);
- Centre for Environment, Fisheries and Aquaculture Science (CEFAS);
- Crown Estate;
- Department for Transport (DfT);
- Department of Environment, Food and Rural Affairs (Defra);

- English nature (EN);
- Environment Agency (EA);
- Greater Thames Archaeological Steering Committee (GTASC);
- Kent & Essex Sea Fisheries Committee (KESFC);
- Port of London Authority (PLA);
- Port of Tilbury London Ltd. (POTLL);
- Royal Society for the Protection of Birds (RSPB);
- Thames Estuary Partnership (TEP);
- Van Oord; and
- Westminster Dredging Company Ltd.

The Group meets three times per year to discuss progress with the PLA's Maintenance Dredging Framework, receive updates on recent and current dredging activity and discuss any issues of relevance to dredging, for example, the Water Framework Directive.

Information Exchange System

An effective planning process requires an up-to-date and relatively detailed knowledge of the environmental sensitivities in an area. One method of achieving such knowledge is through data sharing by key stakeholders. The use of Geographic Information Systems can be invaluable in the presentation and analysis of environmental data. The PLA has developed an Information Exchange System (IES) to hold environmental data relevant to decision-making on dredging applications. Figure 1 is an output from the IES. The IES has been populated with data provided by partners on the Dredging Liaison Group and group members have access to the IES to inform their own decision-making processes.

Data sharing in this manner is an example of best practice and ensures that the PLA has access to up to date information about the environmental sensitivities in a specific area. The need to ensure the data is current is managed by the TEP which has agreed Letters of Commitment with each partner. Data updates occur on a quarterly basis and are incorporated into the IES by the PLA's GIS team.

The IES is used to inform the first step in the application assessment process and will highlight potential environmental sensitivities and act as a trigger for consultation. In this way, if a routine maintenance dredging application is received and the details (quantity, methodology etc.) are the same as in previous years then no consultation will occur. Instead Dredging Liaison Group members will be notified of the start date of the dredge. In contrast, if a new dredging application is received or if an existing operation is to be changed then consultation will be carried out with those bodies that have interests in the area.

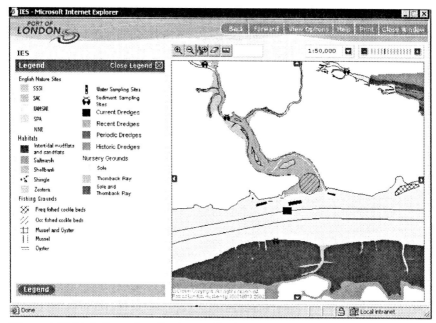

Figure 1 PLA's Information Exchange System

Environmental Considerations

It is inevitable that all maintenance dredging operations will have some effect on the marine environment. Increases in suspended sediment can have an adverse impact in some locations but would be indistinguishable from natural background variations in others. The act of dredging removes (or, at least, relocates) seabed sediments and thus associated seabed habitats and frequent dredging will limit the potential for recolonisation of those areas. It is, however, possible to plan maintenance dredging to minimise environmental impacts and potentially to provide associated environmental benefits, for example, retaining sediment within an estuary system or using dredged material to create or enhance habitats.

Further key considerations in the planning process are the navigation timescale (for example, if a ship is expected at a certain date or, indeed, a deeper draft vessel calls at very short notice) and the economic constraints on a port or harbour. Dredging vessels are often booked in advance or are opportunistically called into a port *en route* to a different location. Ports with limited dredging requirements or budgets may not have the flexibility to define the dredging period.

The first step in the planning process is to characterise the dredge location in terms of the navigation requirements (ensuring that dredging is to be minimised), material type and quantity, dredger availability, disposal options and environmental impacts. The potential environmental impacts are considered in the following paragraphs.

Coastal Processes

The effect of maintenance dredging on the hydrodynamic regime will depend on the depth of sediment to be removed, the frequency, and the sensitivity of the local area.

In general, maintenance dredging is the removal of ongoing sediment deposition and could be considered to be *an intimate part of the sediment regime and dynamics of an area* (UK Marine SAC Project, 1999). Potential effects will also depend upon the selected dredging and disposal methodologies. For example, in a sediment starved estuary, ongoing removal of sediment and disposal to sea or land may well have an effect on the hydrodynamic regime by limiting the sediment supply. In contrast, in an estuary with high sediment input from the sea or rivers, frequent removal of sediment will maintain the status quo.

Port and terminal operators are encouraged to understand the hydrodynamic regime in the vicinity of their dredged area and to plan their dredging methodologies to be sympathetic to the needs of the system. In the Thames, much of the sediment is retained within the system by the use of Water Injection Dredging (WID) and ploughing techniques and is thus available for deposition on the mudflats. Research has suggested that, due to sea level rise, the Thames estuary will require significant and ongoing input of mobilised sediment if the important mudflat habitats are to be conserved (IECS, 1993).

Sediment and Water Quality

An assessment of sediment quality is a key part of the dredging licence assessment process. Licence applicants are required to collect and analyse samples prior to submitting their application. Although there are currently no published UK guidelines on sediment quality, there is an accepted list of parameters for which sediment should be investigated. These parameters include a suite of trace metals, tributyl tin, PAHs, PCBs, microbiological parameters, particle size and total organic carbon. International agreements are encouraging the use of bioassays in assessing the actual effects of deposited sediment on organisms and CEFAS are presently undertaking a bioassay trial in their dredged material assessment process.

In the absence of published UK guidelines, terminal operators can characterise their sediment by referring to sediment quality standards from other countries (i.e. Canadian ISQG or Dutch standards) but these should be used with caution due to differences in geology in other countries and therefore background metal concentrations. In addition, there remains uncertainty about the relationship between sediment quality guidelines for seabed impacts and the levels that would be necessary to prevent water column impacts during the dredging process. It is probable that this discrepancy will be resolved during the implementation of the Water Framework Directive.

In the Port of London, the PLA refers to both the unpublished CEFAS Action Levels and the Canadian ISQG levels, with further consideration about effects on water quality standards dependent upon the level of contamination, dredging methodology and timescale. Land-based quality standards (i.e. CLEA, previously ICRCL) are not considered appropriate for use in assessing effects in a dynamic aquatic environment.

In addition to the potential sediment contamination issues, the PLA will also consider the effects of dredging operations on dissolved oxygen (DO) and suspended sediment levels. Marine organisms, for example fish, are known to be sensitive to changes in DO and very low levels can cause fish kills. This is particularly relevant to the Thames where Thames Water (the water utility) operates the "Thames Bubbler" and the "Thames Vitality", vessels designed to raise dissolved oxygen levels in the water column. Research to date does not suggest a link between dredging and DO levels, but the potential for effect will be dependant on the amount of organic material in the dredged sediment. Applicants for dredging licences should consider if their dredge area is located close to a storm water overflow or sewage outfall.

It is clear, however, that DO levels are greatest during periods of high river flows caused by rainfall in the winter months. Winter dredging would remove to an absolute minimum the potential for an effect on DO levels and could, therefore, be considered as best practice.

Dredging causes disturbance to the material on the seabed and increases the amount of sediment in suspension in the water column. The increase is dependent on the method of dredging employed, for example, a backhoe dredge will release less suspended sediment that a trailer suction hopper dredger. The Thames has naturally high levels of suspended sediment and, in most cases, increases caused by dredging will be within natural variations. However, some species, e.g. shellfish, are particularly vulnerable to increases in suspended sediments. If an applicant intends to dredge in proximity to shellfish or other sensitive species, then consideration should be given to selecting a dredging method that reduces dispersion of suspended sediments.

As a general rule, operators should make themselves aware of the background levels of suspended sediments in the vicinity of their dredge site and should select a dredging method that is compatible with the natural background levels, taking into account seasonal changes, such as levels being higher in winter due to storm events.

Fisheries
The tidal Thames is very important for fisheries and contains 121 species of fish (pers comm. EA, 2003). In particular, the Thames provides spawning and nursery grounds for a variety of fish including sole and bass. Species of conservation interest (e.g. twaite shad) have been observed in the Thames and an extensive salmon restocking programme has attempted since the 1970s.

With the exception of salmon and the conservation species, there is no specific legislation to protect fish resources in England and Wales, but the PLA must consider effects on fish under the reference to conservation of fauna in part (a) of Section 48a of the Harbours Act 1964. In general, fisheries are unlikely to be affected by individual small-scale dredges (unless carried out in a geographically restricted spawning habitat) but there may be cumulative effects from a number of small dredges or from large-scale dredges.

As part of its Maintenance Dredging Framework, the PLA is working with the Environment Agency and Kent and Essex Sea Fisheries Committee to identify any general measures that can be applied to minimise disturbance to fish at sensitive times. To date, the following principle has been agreed for application to dredging projects on the Thames:

1. The area upstream of Tilbury provides nursery habitat for juvenile fish during the summer months. These fish are particularly sensitive to the combined effects of

dredging, outfalls and development during the hot summer months of June to August when dissolved oxygen levels are at their lowest. Where possible and economically viable, dredging operations should be planned to avoid this period and, if feasible, dredging should be programmed to the winter months when biological activity is lowest.

If it is necessary to dredge within this summer period, operators are encouraged to select a dredging technique that will minimise increases above background of suspended solid levels in the surrounding water column.

Further consideration is being given to the sensitivity of sole in the inner and middle estuary to dredging operations during the sole spawning period of March to May and when juvenile sole are present during July to October.

While there is no scientific evidence to demonstrate the direct effects of dredging on fish at these sensitive times, the PLA is taking a precautionary approach in recognising the increased sensitivity during nursery and spawning periods. Adult fish are generally considered able to avoid dredgers and sediment plumes from dredging vessels but juvenile fish and spawning adult fish may be less able to move away from a source of disturbance.

Shellfish

The outer Thames Estuary is a very important and productive shellfishery with the cockle beds being the most extensive. The shellfish are generally found on the drying banks rather than in the subtidal channels. The PLA is responsible for undertaking maintenance dredging in the navigation channels and considers any effects from a dredging plume on adjacent shellfish beds, both in terms of smothering and redistribution of contaminants for potential uptake by the shellfish.

Marine Biology

Dredged areas are periodically disturbed and sediment hosted habitats are removed by the dredging process. This is a direct effect of dredging that it is difficult to mitigate. However, biological productivity is lowest in the winter months and dredging during this period would minimise the effect on biological communities. In addition, prolonging the period between dredging operations would allow the dredged area to play a greater role in the wider ecosystem. Estimates of recovery vary but for muddy dredged areas, the site may be recolonised within a number of weeks.

Birds

Sensitivity from dredging operations is mainly related to over-wintering birds for which many conservation sites are designated. However, when considering the potential impacts the following questions must be asked:

- Will the dredging directly affect any intertidal area?
- Will the dredging activity take place at low tide?
- Will the dredging operation be noisier or cause more disturbance than routine activities at the site? (e.g. cargo handling)

In many cases on the Thames the answer to all the questions is no as the dredging is taking place at commercial jetties in deep-water berth boxes. In these cases, there is unlikely to be an impact on over-wintering birds.

Archaeology
Maintenance dredging operations are removing sediment that has accumulated on a daily basis due to the tidal movements. Maintenance dredging does not involve deepening areas to depths that have not previously existed. There is therefore, low potential for direct impact on sites of archaeological significance. Terminal operators should, however, ensure that they are familiar with archaeologically sensitive sites in the vicinity.

Fishing Activity
There are two aspects to managing the relationship between maintenance dredging and fishing activity. Firstly, fishing vessels often need approach channels and berth areas to be maintained for their own access and should, therefore, have an understanding of the necessity for such dredging. Secondly, the aim of the regulator and operator should be to minimise interference between the two activities. The PLA's IES contains information on the different types of fishing activity carried out in different locations and at different times of the year. This is a further consideration in the decision-making process.

In addition, fishermen are also concerned about effects of dredging operations on juvenile and spawning fish and should be supportive of dredging operations that embrace the best practice of minimising activity within these periods. Clearly there are a number of dredging operations that would have an adverse effect on both fisheries and fishing activity and these should be avoided, for example, using a form of hydrodynamic dredging adjacent to productive, commercial shellfish beds.

Commercial Considerations
In the case of maintenance dredging (and excluding operations within sites designated under the Habitats Directive) the role of the regulator is to achieve the balance between environmental and commercial considerations in the context of sustainable development. Therefore, in addition to the environmental sensitivities outline previously, the following commercial factors should be considered:

- Maintenance dredging is expensive.
- Dredging vessels are often booked in advance.
- Limited supply of dredgers as they work both in the UK and Europe.
- High mobilisation costs for smaller quantities reduced if shared campaign.
- Disposal facility may have specific requirements.
- Need consents to be in place.

These issues are of particular relevance in the consideration of the need for timing and methodology constraints.

Conclusion
Both terminal operators and regulators can ensure that maintenance dredging operations take an appropriate level of account of the sensitivities of the estuarine or

marine environment. Regulators can facilitate the process by the provision of detailed guidance notes or access to environmental data, for example in the form of a GIS. The PLA's experience with the Dredging Liaison Group has shown the value of having a multi-sector group hosted by a neutral body in which ideas can be exchanged and concerns raised.

It is important that constraints on dredging operations are supported by relevant scientific evidence or justified reasoning and that such constraints are considered in the context of other activities that are ongoing in the area.

Best practice in relation to the environmental considerations discussed in the paper can be summarised as follows:

1. Recognise the importance of selecting dredging equipment appropriate to sediment type, contamination levels and background suspended solids levels;
2. Recognise the increased sensitivity for juvenile and spawning fisheries, particularly in combination with pressures caused by other activities; and
3. Recognise the importance of dredged area as potential contribution to ecosystem and dredge in less biologically productive times.

However, these considerations must be balanced with the following commercial factors:

1. Consider dredger availability, mobilisation costs and disposal requirements.
2. Be aware that navigation requirements may not allow flexibility in timing.
3. The sediment characteristics may restrict the choice of dredging methodology.
4. Dredging plant may not be available at the preferred times.

The regulator, in the context of sustainable development, environmental legislation and national and local plans, will balance these issues in order to achieve the desired result of sustainable dredging.

References

ABP Research, 1999. Good practice guidelines for ports and harbours operating within or near UK European marine sites. English Nature, UK Marine SACs Project. pp120.

Institute of Estuarine and Coastal Studies, University of Hull, 1993. The Thames Estuary, Coastal Processes and Conservation.

Port of London Authority, 2004. Maintenance Dredging in the Port of London. In Press.

Port of London Authority, 2004. Planning for Dredging and the Environment on the Tidal Thames. In Press.

Methods of mitigating environmental impact

Anthony Bates, Dredging Consultant, Somerset, UK

Introduction
There is a widespread conception that it is inevitable that any dredging activity will damage the local marine environment. This is true, but only to a very limited extent. Areas of the seabed, or riverbed, from which bottom sediments are removed by dredging, will suffer extensive disturbance, which inevitably will cause temporary damage.

However, in maintenance dredging, the areas of disturbance are very limited in extent, being confined to navigation channels and harbour areas. For example, the dredged channel through Lough Foyle serving the Port of Londonderry in Northern Ireland, plus the Port berths, although 10km long, covers only 0.3% of the total Lough area. In many other UK estuarial ports, including Belfast, Ipswich and Milford Haven, the percentages are also tiny. In many of these sites, recovery following dredging will be rapid (Bamber, 1980) and the long-term effects may not be significant. Regardless of any localised impact, if channels and harbour areas are to be maintained in a condition that is safe for navigation, occasional disturbance is inevitable.

Concern should therefore be directed to areas beyond the immediate limits of channels and berths, where if sediments are widely dispersed in suspension, the extent and consequences of any impact may be more significant. It is therefore those areas outside of those maintained by dredging where it is most important to control and limit the impact of dredging.

This paper describes measures that can be adopted to control the impact of dredging on areas adjacent to dredging, or within the areas of potential impact.

However, before doing so, it important to recognise that maintenance dredging, even if poorly controlled, will rarely have as great an effect as natural events, such as storms, or floods. Nor will the effect be as great, or as widespread, as that commonly caused by commercial fishing operations.

In 2001 in UK waters alone, commercial fishing removed in excess of 300,000 tonnes of fish and shellfish. It has been estimated that the tonnage killed by fishing gear, or as by-catch, may be as high as 14 times the recorded tonnage landed. A large beam trawler may severely disturb 400 hectares of seabed in a single day. The extent of areas affected by beam trawling in sea areas around the UK totals 1.25 million km^2. Some areas may be fished 4 times a year. In comparison, the area of seabed disturbed by maintenance dredging in the UK totals approximately 35km^2, or only 0.003% of that adversely affected by fishing.

Alongside the large-scale disturbance by fishing, the impact, even of poorly controlled dredging, might reasonably be considered to be insignificant. Is it necessary therefore to control dredging operations? The answer of course is yes, partly because, whilst some aspects of modern commercial fishing may be questionable, appropriate fishing is beneficial, at least for mankind. Furthermore, the marine environment within which large scale commercial fishing takes place is different to the coastal and estuarial environments where maintenance dredging is necessary. Appropriate controls should therefore be employed such that the relatively insignificant impact of dredging may be made so insignificant as to be virtually harmless.

Cause and effect

Because most maintenance dredging today, probably in excess of 90% by volume in Europe, is carried out by Trailer Suction Hopper Dredgers (TSHD), this paper deals primarily with controls that can be applied to this method of dredging, but first, it is necessary to consider the ways in which the dredging activity and processes have potential to cause adverse impact on the marine environment.

Toxic pollution

Pollution may result when contaminated sediments are disturbed and dispersed. Dispersion is considered below. It is unusual in UK waters for sediments to contain significant contamination in areas that are dredged regularly, but where significant contamination is present, special dredging methods will be necessary. This is a separate subject and is not addressed here.

Pollution may also result from leakage from the dredger systems, such as hydraulic, lubricating, or fuel oils. However, this can occur with any vessel, dredger, or otherwise, and avoidance is largely a matter of appropriate maintenance and regular inspection. For some types of dredger, such as Hydraulic Backhoes, where if leakage occurs it may be uncontained, it is good practice to specify the use of vegetable oils in hydraulic circuits.

Smothering

If large volumes of sediment are released into suspension, dispersed and settle in significant thickness onto sensitive and immobile receptors, such as shellfish beds, it is possible that damage will be caused. Where such risk exists, it is necessary that measures be taken to prevent the large-scale release and dispersion of sediment. However, this may not be important at every site. For example, the Severn Estuary in the UK is so dynamic and the natural level of sediment suspension at times is so high that the added effect of dredging is unlikely to be significant.

Turbidity

If fine sediments are released into the water column and remain in suspension for a significant time, fish may be damaged (Sigler, 1990; Pennekamp, 1990), or may be deterred from entering the area. When deterioration in water clarity is likely to have an adverse effect on fish health, or passage, measures should be taken to control the rate of release and dispersion of fines.

Noise

For most maintenance dredging work, the sediments to be dredged are fine grained, usually clay, silt, or sand. When a TSHD is employed to dredge these materials, noise levels will be similar to those generated by normal commercial shipping and hence noise is unlikely to create any special problem.

For some types of stationary dredger, such as bucket ladder and backhoe types, noise levels may be such that work at night has to be restricted if close to inhabited areas, or sleep may be disturbed. Noise, or vibration created by some stationary dredgers could also have an effect on local fisheries. The noise levels generated by dredging plant are unlikely to be so high that fish may be damaged, but in specific circumstances, might be sufficient to deter fish from entering the area and hence disrupt feeding, or the migration of some species.

Overall periods of maintenance dredging are usually quite short and rarely continuous. For TSHD's, there will be periods when the dredger is off-site sailing to and from the disposal area. These periods may be longer than that occupied in dredging. There will therefore almost invariably be periods when the feeding or migration of fish is unaffected. Furthermore, it is most unlikely that the effect of local dredging will extend across the whole water section, or indeed, even across most of the water section. There will therefore usually be substantial areas of clear water where the behaviour of fish is unaffected.

Dredging methods

Trailing Suction Hopper Dredger (TSHD)

Most readers are probably familiar with the general design and mode of operation of a TSHD. However, for the benefit of those who are not, a brief description follows.

The general outward appearance of a TSHD is similar to that of a commercial ship. See Figure.1. It may be quite small, perhaps only 60m in length with the capacity to carry only a few hundred cubic metres of dredged mixture, or quite large, currently up to 230m long and carrying up to 33,000 cubic metres ("Vasco Da Gama").

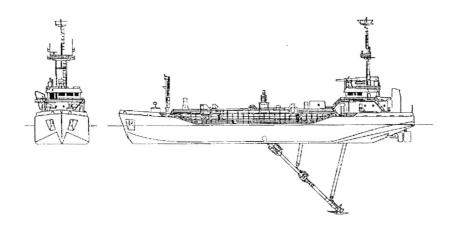

Figure.1 – Small Trailer Suction Hopper Dredger

Dredging takes place with the dredger moving slowly ahead under it's own power, typically at a speed of about 2 knots (4km/hr). By design, the ship is highly manoeuvrable, typically with twin-screw propulsion and powerful bow thrusters.

Over one, or both sides of the vessel, long articulated steel pipes are suspended. These can be raised, or lowered. The upper end of each pipe is connected to a centrifugal pump, usually mounted inside the hull. At the lower end of the pipe, a fabricated steel structure, known as a 'Draghead', is attached. This is broadly similar in shape and function to the suction head attachment on a cylinder type domestic vacuum cleaner. To begin dredging, the draghead is lowered into contact with the river, or seabed.

Steel teeth and high-pressure water jets may be attached to the underside of the draghead to breakout and fluidise the seabed soil. The disturbed soil is drawn into the draghead by suction and transported via the dredge pump to a distribution system of pipes and channels to discharge within the hopper of the vessel.

Depending on the elapsed time of the loading cycle, all, or part of the soil pumped into the hopper, will settle from suspension, adding to the cargo of sediment contained in the hopper.

Surplus water, if any, flows over an adjustable weir, or weirs, usually, at least on modern designs, to discharge back into the surrounding water at a level below the outside water level.

This process of hopper overflow, sometimes referred to as 'weiring', may continue until the hopper contains the maximum design load, or until such time as further loading becomes uneconomic, or causes excessive loss of sediment. The optimum loading time is dependent on the characteristics of the seabed soils and the sailing distance between the point of loading and the point at which the cargo is to be discharged. In fine grain materials, including clay and silt, loading may cease after as little as 15 to 20 minutes, or in sand and gravel, loading might continue for 90 minutes, or more.

If travelling continuously in one direction at an average of 2 knots, a TSHD dredger would cover approximately 3.7 kilometres of channel during a loading time of one hour, but may partially load over a shorter distance, perhaps 2km, and then turn to complete loading on one or more additional runs. If dredging a restricted area, such as a berth, or isolated shoal, trail length may be much shorter. Time spent turning, or going astern, may then add substantially to the total loading time.

Sediment losses during dredging
Apart from the total disturbance of the benthos in those areas that are to be dredged, which cannot be avoided, the most important influence of dredging on the marine environment will be the suspension and dispersion of fine sediments. The suspension of fines is strongly dependent on the characteristics of the soils to be dredged. Usually, coarse particles, such as sand, will quickly settle within, or close to the area of dredging, but fine grain particles, including silt, or clay, may disperse further a field.

Soil characteristics
Seabed soils dredged to maintain, or restore depth, may vary greatly in their characteristics. Soils are characterised by the size distribution of the individual soil particles.

In British Standard terms (BS:1377), soil particles with a dimension greater than 2mm are termed 'gravel', those with size in the range 0.063mm to 2mm are termed 'sand', those in the range 0.002 to 0.063mm are 'silt' and anything smaller is 'clay'.

Usually, almost all soil to be dredged for maintenance purposes will fall within these ranges. Most will be a mixture of clay, silt and sand.

Types of sediment loss
Sediment losses can be divided into two types. Initially, sediment may be dislodged, or disturbed, by the action of draghead, bucket, or cutterhead, but not picked up. This is more correctly termed 'spill'. Usually, most of the spilt sediment will fall back into the immediate area of dredging and hence is not truly 'lost'. This is particularly true for the coarser grain materials, gravel and sand. Sediment that is spilt is most likely either to be dredged on a subsequent pass of the dredger, or to consolidate and remain in place, but a small proportion may be eroded and suspended during subsequent periods of high tidal flow and transported away from the immediate area of dredging.

The alternative type of loss is that when sediment is forced into suspension and remains suspended within the water column for a significant time. Sediment in suspension will be carried away from the working area by the prevailing current. The direction and distance of this dispersion is dependent on many things, but mainly on the settling characteristics of the sediment and the strength, direction and duration of the prevailing current. Sediment dispersion is discussed in the paper of Dr. Michael Dearnaley.

The various ways in which a TSHD might cause the suspension of sediment are described below.

Mechanisms of sediment suspension

Trailing Suction Hopper Dredger
Draghead water-jets
Sediment may be forced into suspension by high-pressure water jets in the underside of the draghead (Figure.2). The rate of suspension is highly dependent on the soil properties. If the soil is loose and fine grain, as with silt, or silty fine sand, then excessive use of water jets may cause excessive suspension. However, sediment that is suspended in this way is unlikely to be forced very high into the water column (See Fig.3) and hence is unlikely to be transported very far, except within the confines of a dredged channel. Such losses cannot be separately quantified, but subject to water jets being used only as and where appropriate, in my opinion the loss will usually be very small.

Propeller scour
Sediment may also be forced into suspension by propeller scour. During dredging the ships propulsion must overcome the resistance of the dragheads trailing through the seabed soils. In some conditions propeller thrust may be quite high. However, scour is only likely to be significant when dredging with very small clearance underkeel, a situation that usually can be controlled.

De-gassing
If dredging organic soils containing gas, it is usual to use de-gassing equipment, the function of which is to avoid cavitation in the dredge pump, with the attendant risk of damage, but also to improve the concentration of solids in the pumped mixture. The process of gas removal includes the ejection of a water/gas mixture that may also contain a high concentration of fines. Usually the mixture is discharged into the hopper of the dredger, but on some older TSHD's, discharge may be directly overboard. However, although the

concentration of fine sediment in the ejected mixture may be high and unsightly, the rate of flow is small. The potential impact is therefore also small.

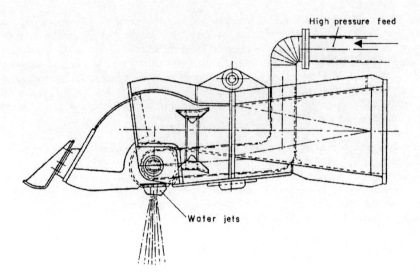

Figure.2 - Draghead water jets

Hopper overflow (weiring)
If loading continues beyond the time when the hopper of the dredger is full of pumped mixture, the hopper will begin to overflow via an overflow weir, or weirs. Sediment will be suspended in water overflowing the hopper weir during loading, particularly during extended periods of overflow, when the trapping efficiency of the hopper will be progressively reduced. This is potentially the most significant source of sediment loss. However, the rate and duration of loss can be controlled by adjusting the flow rate of the incoming mixture. Flow rate can be reduced by reducing the pump speed, or on TSHD's with two suction pipes, can be halved by using only one pipe.

On modern TSHD's the overflow water usually discharges below the outside water level. The operator can and usually will adjust the height of the overflow weirs during loading to maximize the percentage of sediment retained in the hopper. Submerged discharge of the overflow mixture has the advantage of release closer to the seabed, with reduced risk of dispersion beyond the immediate area of dredging. When dredging to maintain incised navigation channels, the submerged sediment plume may not disperse beyond the channel side slopes (see Fig.3). However, a potential disadvantage of submerged overflow discharge is that turbulence may be injected into the submerged plume by the action of the ship propellers. The plume may then disperse more widely.

Loading of the hopper can be stopped immediately at the commencement of overflow, at which time the proportion of solids contained in the hopper will be only the average proportion of solids contained in the pumped mixture from the start of loading until the start of overflow. Many things, the most important of which is the character of the seabed soil, will influence the proportion of solids loaded. The proportion (by volume) of in-situ soils loaded may typically range from 20% to 50%, the latter being typical of loose silts and silty sands. In very weak fine grain deposits, such as fluid, or low-density mud, the concentration of the pumped mixture reach 100%.

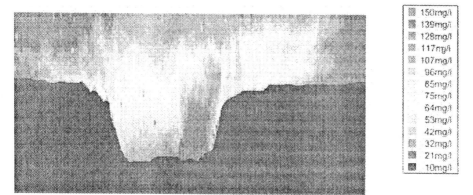

Figure.3 – ADCP image of sediment plume within dredged channel

When the average concentration of the pumped mixture is relatively low, say 20%, it may be uneconomic to cease loading when the hopper begins to overflow, particularly if the sailing distance to the discharge location is long. The alternative is to continue loading beyond the commencement time of hopper overflow, with the object of increasing the proportion of solids in the hopper by retaining the coarser fraction of the pumped mixture and losing some of the fine fraction in the overflow. This is normal practice when dredging for aggregates, or to improve the quality of granular material to be used in land reclamation. Losses are then dependent on the particle size distribution of the dredged soil and the total overflow time. During prolonged overflow, a proportion of coarse sediment will settle in the hopper and a proportion of fine sediment will be lost in the overflow water. As loading continues, the proportion of sediment lost will increase; eventually reaching 100% loss as the hopper load nears the maximum, or the trapping efficiency is zero.

Light mixture overboard
During the loading period pumped mixture may be diverted overboard to discharge above, or within the water column (LMOB). The object of diversion is to minimise the volume of water loaded to the dredger hopper and usually occurs only if the solid content in the pumped mixture is very low (light mixture). On most TSHD's, light mixture can be diverted overboard automatically. With automatic systems (ALMOB), concentration meters monitor the density of the incoming pumped mixture of water and soil. When density falls below a pre-set level, valves are actuated to close flow to the hopper and divert flow to discharge externally. Alternatively, the operator can divert the flow manually.

If sediment loss is to be controlled, it is necessary to limit the use of overboard discharge. Specifying the maximum density of mixture and duration of discharge can do this. The specified limits will be dependent on the characteristics of the dredged materials and the site of dredging. Density may be variable within an estuary, but if assuming a water density of $1.03t/m^3$, it would clearly be unrealistic to specify a maximum limit of 1.03 on the overboard discharge. In this situation, a figure of 1.05 t/m^3 may be appropriate, but although the solids content at this level may be modest, the mixture may be highly coloured, resulting in a clearly visible plume.

The permitted duration of overboard discharge is also dependent on the nature of the dredging task. If the extent of the area to be dredged is very limited and hence trail length is short, hopper loading will not be continuous, but will be regularly interrupted by the need to turn the dredger, or go astern, to begin each new trail run across the area. Pumping to the hopper will cease during each non-dredging period. Pump discharge may then be diverted overboard. The discharge is likely to create a temporary turbid plume, but although unsightly, the concentration of solids in the plume will usually be very low. Alternatively, the pump/s must be stopped and restarted at the end and beginning of each run. At the beginning of pumping the flow will be only water, or a mixture with only very low solids content. Usually this flow will also be diverted overboard.

Intermittent dredging activity

Sediment plumes generated during loading will be interspersed by clear periods when the dredger is off-station sailing to the disposal point, discharging and returning to the area of dredging. These clear periods will be long when the area dredging is a long distance from the disposal area. Typically, when loaded, a small to medium size TSHD may sail at a speed of about 10 knots. Thus the dredger will be absent from the area of dredging for the time it takes to discharge the hopper, usually 5 to 10 minutes, plus about 6.5 minutes for each kilometer distance between the dredge area and the discharge area.

Control methods

There are various methods by which the dredge operator can control the loading process and any attendant losses. These include adjustments to: trail speed (speed of the vessel over the ground), pump flow rate, water jet flow, draghead attitude and down-force, overflow weir level, pump discharge distribution, LMOB and de-gassing. All of these functions are continuously monitored and controlled by the operator from a bridge control station, which on a modern TSHD is very sophisticated.

Sediment suspension during hopper overflow

It has been previously suggested that it is probable that more than 90% of maintenance dredging in Northern Europe is carried out by TSHD. It is also probable in my opinion that more than 90% of all sediment released in suspension during any maintenance dredging operation by TSHD is released during hopper overflow. This of course assumes that some period of overflow is permitted. It is therefore worth considering the effect of hopper overflow in more detail.

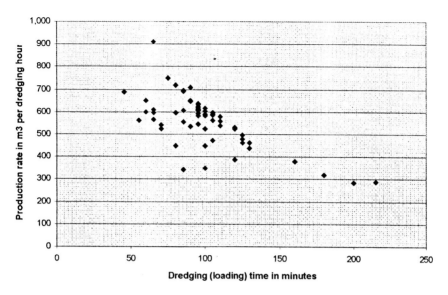

Figure.4 - TSHD production rate versus loading time

It is not uncommon to observe that the Master of a TSHD will tend to continue hopper overflow for longer than is necessary to maximize production. This is probably because the Master instinctively wishes to sail to the disposal area with the maximum possible cargo, even though the maximum achievable cargo may be more than the economic maximum. Figure.4 illustrates the effect of loading time on production rate when dredging fine sometimes silty sand with a small TSHD. Because of scatter in the data, it is difficult to determine a precise 'optimum' loading time, but the trend, of a reducing production rate with increasing loading time, is quite clear. In this example, the time at which hopper overflow commenced was variable, generally ranging between 30 and 40 minutes. The production rate begins to decline after between 60 and 90 minutes. Thus, overflowing by more than an average of about 40 minutes has no economic benefit. The dredged material is fine sand and measurement of the work was 'in the hopper'.

However, had measurement been *in-situ*, by comparison of pre-dredge and post-dredge surveys, extended overflow may have been beneficial, at least for the dredge operator. This because some of the sediment lost in overflow will disperse beyond the limits of measurement and hence count as volume removed. The total volume lost during prolonged overflow, even for a small dredger, may be substantial. In the small dredger example illustrated by Fig.4, the total loss after 85 minutes loading would have been about 175 cubic metres per loading cycle. For an average size (5,000m3 to 6,000m3 hopper capacity) twin pipe TSHD, the loss in similar circumstances would be about 2,800m3 per cycle. However, much of this quantity may quickly settle within the dredging area, either remaining there, or being re-dredged during subsequent passes, and hence may not be true loss. It should be noted that the volumetric losses listed above are *in-situ* cubic metres, each of which may contain a very high proportion of water. Loss would be more accurately expressed in 'tonnes dry solids' (tds), but this is highly dependent on the soil density *in-situ* and is not easily measured.

It is clear that whilst there is potential for overflow losses to be very high, the losses may only become important if the suspends sediments disperse beyond the limits of the dredging area and settle onto sensitive areas, such as shellfish beds, or marine flora. It is therefore important to determine the fate of any sediment plume created by the dredging process.

Measuring dispersion

Until recently it was not possible to easily monitor the progress of a sediment plume, nor to estimate the plume content, with reasonable confidence. This is now possible by using backscatter with Accoustic Doppler Current Profilers (ADCP's). It is not the purpose of this paper to discuss the detail of the use of ADCP's methods. This may be found in Land,J.M. and Bray,R.N. (1998). The object here is to consider how ADCP methods may be used to monitor and control the dredging process.

Usually, the primary objective is to determine the pattern of dispersion of the plume. This can be done quickly and at modest cost by sailing successive tracks around the dredger and through the sediment plume during loading.

The most common situation where measurement is important is that when dredging to maintain navigation channels that pass though, or close to sensitive areas. In such cases the extent of the sediment plume and relative concentration of suspended sediments within the plume can be seen in real time by sailing tracks parallel to the channel. The plume formation can be seen on a VDU as the survey vessel progresses. Data can be logged for later analysis. If a succession of tracks are sailed to monitor plume generation and dispersion at various stages of the loading process and under different tidal conditions, it will be possible to assess whether or not the sediment plume is likely to affect any area that is of concern.

The plume tracking exercise should be carried out over a representative number of cycles at the commencement of a dredging campaign. It is useful during the initial work to deliberately exaggerate the suspension of solids by overflowing the hopper during periods of high solids concentration in the pumped mixture, such that the sediment plume created is more severe than will occur during normal operation. If the plume characteristics are such as to be considered harmless this will give confidence that normal operation is unlikely to have adverse effect.

The initial plume Tracking will ideally be timed to coincide with Spring tides, such that the extent of plume dispersion and hence the potential impact on far field sensitive receptors is maximized. The quantity of sediment released during these initial few cycles, even if transported to sensitive areas, is unlikely to be sufficient to cause harm. Conversely, if sensitive areas are close to the area of dredging, it may be preferable to conduct trails during neap tides, when the plume is less likely to be diluted by widespread dispersion.

CONTROL METHODS & (Relative importance)

The following paragraphs describe the methods by which the suspension of bed materials during dredging may be controlled. The level of control necessary is dependent on the sensitivity of the local marine environment to the suspension, or deposition of fine sediments. Control will not be necessary at all sites.

The relative importance attached to each potential cause of sediment suspension is marked out of 10, 10 being the most significant potential cause of suspension. The relative score is subjective and will not be applicable to every situation.

Propeller scour (2)

This is unlikely to be a problem unless underkeel clearance is small and the bed materials are loose and fine grain. Control is unlikely to be necessary where bed materials are sand, or gravel.

Underkeel clearance can be increased by employing a dredger with shallow draught, or by restricting operation to avoid periods when tide height is not sufficient to provide adequate clearance.

If seabed materials offer significant resistance to trailing, it may be necessary to reduce draghead penetration to reduce trailing resistance and hence also reduce the propulsion effort necessary to maintain trailing speed.

The low relative importance of (2) reflects the fact that scour is only likely to occur under the narrow range of conditions described and because any suspension of sediments initially occurs at the bottom of the water column, thus minimising the risk of significant dispersion beyond the area of dredging.

Draghead disturbance (1)

This is unlikely to be a problem unless dredging is in very fine cohesionless soils in a very sensitive environment.

Reducing draghead penetration and trailing speed will reduce disturbance.

Water jets (3)

It is unnecessary to restrict water jet use if dredging in soils comprising only sand and gravel size particles.

If dredging in fine grain sediments, the use of water-jets may not be required. However, in some such soils production might be increased when the use of water-jets has the effect of fluidising the soils, or changing the soil density, such that pump performance is improved. It may then be necessary to achieve a balance between production and sediment suspension.

Suspension may be reduced by increasing draghead penetration, or reducing water-jet flow.

It is possible, though unlikely, that in very sensitive environments it may be necessary to avoid the use of water-jets completely, but this could result in reduced production and hence increased cost.

Light mixture overboard (LMO)(4 to 10)

In most situations, at least limited use of 'Light mixture overboard' (LMO) will be necessary to achieve an economic rate of production.

Unlimited, or continuous use will amount to agitation dredging and hence should not be applied in a sensitive area.

In most cases it will be appropriate to restrict both the maximum total duration of discharge overboard and the maximum density of mixture discharged. The level of restriction necessary, if any, should be determined by local circumstances.

Hopper overflow (4 to 10)
As in the case of LMO, continuous overflow is agitation dredging. In fine grain sediments, clay, silt and very fine sand, hopper overflow is unlikely to result in increased production, unless the effect of agitation and dispersion is included. In coarse grain sediments, the opposite is true. The production rate is likely to be low unless some period of hopper overflow is permitted.

The effect of hopper overflow can be estimated by comparing the rate of solids pumped with the gain in hopper load. On modern TSHD's the rate of solids pumped and the change in hopper load is measured by onboard instrumentation. The difference over any period of time will be the volume, or weight, of sediment lost in suspension.

When dredging in medium to coarse sand, or gravel, sediment lost in overflow is likely to descend rapidly through the water column and settle within the areas of dredging, or close by and hence is unlikely to cause concern.

With decreasing particle size and hence lower settling velocity sediments may disperse more widely and it may be important to limit the rate and total quantity released. The level of restriction will be dependent on the actual sediment characteristics and the sensitivity of the local marine environment. It may not be unreasonable to specify zero overflow time if the average particle size is less than 0.075mm. To maximise production when loading is to stop at the point of hopper overflow, the hopper should be completely empty when loading commences.

Conclusions
The impact of maintenance dredging on the marine environment within which the activity takes place can be controlled by various methods, such that the impact beyond the immediate limits of dredging can be reduced to harmless levels.

The most important tool for use in maintenance dredging is the trailing suction hopper dredger, which probably accounts for more than 90% by volume of all maintenance dredging in Northern Europe. The most effective controls when using a dredger of this type are those placed on hopper overflow and light mixture discharge during loading.

Controls will not be necessary in all situations and need not and probably should not be imposed, unless the effects of dredging are likely to be greater than those that regularly occur naturally.

When controls are found to be necessary, the level of control, in terms of the operational restrictions imposed, should be determined relative to the local marine environment.

Any restriction on the operation of dredging plant in most cases will result in reduced production and hence increased cost.

References

Pennekamp, J.G.S. & Quaak, M.P. 1990. Impact on the environment of turbidity caused by dredging. *Terra et Aqua*. No.42, pp 10-20, International Association of Dredging Companies (IADC), The Nethrlands.

Sigler, JW 1990 Effects of chronic turbidity on anadromous salmonoids: Recent studies and assessment techniques perspective. In effects of dredging on anadromous pacific coast fishes. Workshop proceedings, Seattle, September 8-9, 1998. Pbl:Washington Sea Grant Programme/Univ.of Washington, Seattle.

Treatment and beneficial re-use of contaminated sediments

Guy S. Pomphrey, UK Manager, DEME Environmental Contractors NV, East Grinstead, UK
Geert Van Dessel, Technical & Commercial Executive, DEME Environmental Contractors NV, Zwijndrecht, Belgium

Abstract
Port development, maintenance and natural processes alter the hydrodynamic conditions of waterways, resulting in accumulation of sediments in unwanted areas. Unfortunately, these sediments are often (historically) contaminated by point and diffuse sources upstream. In the last decade, several environmentally acceptable and sustainable treatment techniques for contaminated dredged material (CDM) have been developed and put into practice on a large scale. These include dewatering and separation techniques, bioremediation and both in- and ex-situ immobilisation. This paper will describe some of these processes, and also the issues of fixed and mobile treatment centres and beneficial re-use options.

Introduction
The sediments encountered often contain increased levels of heavy metals, PAH's, PCB's, mineral oil, fuel and other organic or inorganic residues. Industry, as well as agriculture and human life are responsible for the spreading of these pollutants. In some cases it concerns very specific compounds like Tributyl-Tin (TBT), which not only has shown disastrous effects on the aquatic bio-fauna, but also necessitates sophisticated and specifically monitored treatment.

Since national, regional and European legislation have become ever more stringent, contaminated sediments are considered waste products. Therefore they may no longer be freely distributed on land or used in agriculture, and dumping at sea is restricted. Two options are left open to deal with this flow of "waste products": landfilling, e.g. sub-aquatic disposal or on land in confined disposal sites, or treatment, in order to recycle and beneficially reuse as much of the material as possible.

Disposal is the easiest solution but is not environmentally sustainable and leaves a heritage for the future. At the same time, environmental taxes and limiting acceptance criteria on the dumping of waste are increasing, and proper management and controls for landfills are becoming more complicated. Some authorities reward or even require the re-use of dredged material in capital works. They take a more risk-based approach to re-use, instead of general standards based on concentrations. It is clear that pollutants only pose a threat to living organisms and the environment, when they can leach out of the material, and taken up by the environment. Many techniques of stabilisation and bioremediation form end products that are perfectly safe to re-use, thereby reducing the cost of treatment.

Treatment techniques
The first step of all treatment procedures for dredged sediments is dewatering. The evacuation of excess water minimises the volume of material to be handled, mixed or treated. Furthermore the processing of the aqueous fraction is in most cases less complicated.

Dewatering
There are two main ways to dewater sediments: accelerated natural dewatering in treatment lagoons, also know as "lagooning", which is an enhanced drainage and filtration technique, and mechanical dewatering. Practical experience has shown that both techniques should be applied under different conditions. The soil texture of the sediments is one of the main parameters in the selection process. Sand as a pure product is often dredged and re-used as such.

As a general rule the more sandy or silty the material, the more efficient the lagooning processes. The pore structure enhances the removal and transport of the water in a natural way. Sand and silty materials are more abrasive for mechanical dewatering equipment. The more clayey the sediments, the less efficient the lagooning. Without any action, the clays will remain expanded, thus not releasing their water.

Also the presence of pollutants must be considered. Lagooning is an aerobic process that stimulates biological activity. By enhancing this process, bioremediation effectively decontaminates sediments during the lagooning process.

Dewatering costs per ton dry solids decrease with increasing dry matter content, with mechanical dewatering becoming cheaper than lagooning for very low initial dry matter contents (15-25%).

Sand Separation
When 50% or more of the dredged material represents sandy particles, sand separation is a must. Most pollutants have the tendency to be bonded to the finer particles. This means that, when the (clean) sand is separated out, the total volume of material that needs to be treated is greatly reduced. At the same time, sand has excellent re-use potential, e.g. in the construction industry, which will help to cover the costs of the treatment.

Different methods of sand separation are used, like hydrocyclones, for example, or settlement ponds. Both make use of gravity to separate the heavier sand particles from the lighter silt or clay particles. Sieving and washing make it possible to separate specific particle sizes.

Accelerated natural dewatering ("lagooning")
In case large volumes of sediments need to be treated, it has been established that the most efficient way in terms of productivity is the lagooning technique. The dredged sediments are brought into large lagoons, exposed to the elements, where the water has time to evaporate, or is evacuated through the system of drainage tubes. Typically, the lagoon is filled up by a layer of 1 to 2m thickness. Every time a crust forms on the sediments, the top layer is broken, and the whole pile is tilled. The efficiency of lagooning is governed by the following

(generally known) major factors: the weather (i.e. sunshine, rainfall and wind), the layer thickness, the time between tilling, the dry matter content and composition of the incoming material, etc.

Each type of sediment behaves differently, basically depending on the soil characteristics of the material, like grain size distribution, organic matter content, presence of some chemical compounds, etc.

At the start of the process, the drier the material, the faster it will reach the point where it will start to behave as a soil, at a density level of 1,5 to 1,6 ton/m³. Sediments entered at various density levels can be processed. Process-time to obtain predefined density levels has been measured. Figure 1 clearly shows the increase in time needed when sludge is diluted when brought into the dewatering lagoon.

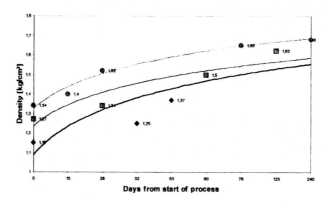

Figure 1: Lagooning of sediments at various density levels

The time required to obtain final soil structure is largely inversely proportional to the number of times the material is tilled during the process. The amount of energy used in the system will therefore have a large influence on the efficiency of the lagooning process.

The water that comes out of the dredged sediments may contain some mobilised pollutants. The water is collected in the drainage system, from where it is pumped to a wastewater treatment plant. Here the pollutants are degraded or adsorbed.

Mechanical dewatering

Historically, filter presses, used for mechanical dewatering, have shown low productions. Capacities of only up to 1 ton of dry solids per hour per cubic meter of filter press volume were achieved. In addition, the classical concept of a filter press does not allow any coarse materials (sand or fine gravel) in the sludge, making an extensive preliminary screening and sand separation necessary. Unfortunately, this sand fraction would otherwise help to speed up the dewatering and increase the filter cake's geo-technical properties afterwards.

In recent years however, mechanical dewatering with filter presses has again gained interest and is to be implemented on a large scale (e.g. Metha-plant Hamburg, Port of Antwerp). At the same time, mobile dewatering units, based on filter presses, have recently been used in some dredging projects (e.g. Belfast, Port of Brussels, Canal Brussels-Charleroi). The basis for these mobile plants is a compact filter press of about 2,8 m^3 cake volume. A combination of a robust piston-membrane pump, parallel feeding of the filter press chambers and wear resistant filter cloth, now allows sand and fine gravel particles up to 10 mm and enables productions of up to 3 tons of dry solids per hour per cubic meter of filter press volume. The high tolerance with respect to grain size reduces the need for pre-treatment to a simple screening and sieving process, and prevents dilution of the sludge with water, which would only decrease the production drastically. The dewatered sludge cakes contain between 65 and 70% of dry solids.

The process is fully automated and can be controlled by a single operator and an excavator driver. A laboratory, including a small filter press, is available, making lab scale dewatering tests possible, as well as determining the exact doses of flocculants to be added. In addition, a generator makes the installation fully self-supporting.

The process consists of several stages, the first stage being the loading. As dredged sludge from inland waterways often contain considerable amounts of stones, debris and waste, the unloading excavator picks out the coarse waste (bicycles, refrigerators, etc.) from the barge and stores this in a separate container on deck of the pontoon. Most inland waterways are dredged mechanically, i.e. by a bucket dredger or by a pontoon-operated excavator. This means that the sludge arrives at around 40 to 50 % dry solids.

In step two, the pre-screening stage, the coarser fraction of the sludge, consisting predominantly of waste and debris, is removed. In order to achieve this without adding a substantial amount of water (which should only lower the dewatering capacity of the filter press afterwards) the concept of a rotating drum was chosen. A rotating drum breaks the cohesion in the clayey sludge, screens out the coarse material, and washes this waste at the end of the drum with a limited amount of water. By washing the waste, the landfill cost for this fraction is significantly reduced.

In order to limit abrasion of the filter cloth, debris larger than 4 mm is not allowed into the filter press. A shaking sieve at 4 mm removes the fraction 4 – 10 or 4 – 20 mm. The oversize of this sieve is again washed on the sieve by adding a minimum of water. In most cases, this fraction consists predominantly of gravel, which can be reused without further treatment. If the material should contain too much waste or organic material, it is further separated in one of DEC's soil treatment facilities.

The buffer container underneath the shaking sieve can store up to 25 m^3 of sieved sludge. The sludge is now at a dry matter content of about 35 %, which is ideal for dewatering by the filter press. Prior to dewatering, the sludge is conditioned by the addition of lime, the third step. This happens in the two conditioning tanks, which feed the filter press pump.

The use of lime, to improve dewatering, has two additional beneficial effects on the filter cakes that are produced. First of all, the lime improves the geotechnical properties of the cakes; higher levels of dry matter have to be reached to obtain the same geotechnical quality for lagooned sludge. Secondly, part of the organic pollutants and most of the heavy metals are immobilized, creating a new reuse potential for lightly polluted sludge.

Finally, in the last stage, the sludge is dewatered by a quick filter press, which has a capacity of up to 6.5 tons of dry solids an hour (about 30 m³/h at the input of the filter press or 20 m³/h of sludge at the input of screening drum) or up to 10 tons of filter cake an hour. This high capacity is not only due to the fact that the sand fraction (even when very small) is allowed in the press (which increases the effectiveness of the process), but also due to the parallel feeding of the chambers in the press.

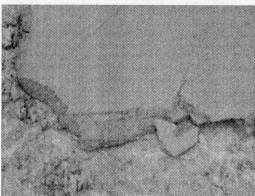

As the screening, sieving and conditioning equipment is designed for 60 m³/h of sludge, the dewatering step can be extended by just adding two extra filter presses. This will increase the capacity of the whole installation to 20 tons of dry solids an hour (up to 30 tons of filter cakes an hour). Recently, DEC have completed a project in Belfast, using this same dewatering installation, installed on shore, close to the dredging operation. About 12.000m³ of sediments from the river Lagan were dredged and effectively dewatered.

Bioremediation
Bioremediation occurs at a very slow rate, since the level of oxygen available for the indigenous micro flora controls it. Sediment on the bottom of rivers is in a near-anaerobic state, which causes the black colour associated with it. This colour is the result of a chemical reaction, which depletes the sediment of oxygen, e.g. sulphate to hydrogen sulphide. The anaerobic conditions in the sediment together with the small particle size inhibit normal indigenous micro-flora, therefore making natural attenuation impossible. However, with the right pre-treatment, it is possible to return to natural conditions, boosting bioremediation. To enhance bioremediation, organic fertilising materials together with soil conditioners are applied to the dewatered sediments.

Since the sediment was tilled during dewatering, the material will have started to crumble, which improves the aeration. Nutrients for the indigenous micro-flora are supplied to the sediment-compost. This also improves the overall grain size. Beneficial effects are further obtained by using liming techniques. Apart from the improvement of the overall grain-size distribution and formation of stable aggregates the high permeability ensures further rapid dewatering and oxidation. This is due to the fine pores created by the lime, which results in high porosity even when compacted.

After sufficient mixing the sediment is placed in ridges for optimal bioremediation. Moisture content of the sediment is monitored to ensure proper remediation conditions. Excess rainwater runs off the ridges and seeps into the drainage system. Ridges offer several advantages. They reduce the horizontal surface exposed to the rain and create a larger vertical surface for further wind evaporation, and solar contact. Every week the ridges are turned and aerated using specially designed windrowing equipment, to supply the sediment with necessary oxygen. During this process remaining lumps of sediment/sludge are also crushed. This results in a larger surface area and hence faster bioremediation.

Immobilisation

Often, the physical and/or chemical properties of the sediments need to be changed in such a way that the spreading of the contaminants by leaching, erosion or dispersion is significantly decreased. In that case an immobilisation technique is required.

Contaminants can only cause an environmental hazard if they are available for take up by living organisms. Through immobilisation of the contaminants, those risks are reduced or eliminated. By combining the mixing of the sediments with immobilisation additives and stabilising (puzzolanic) materials, the physical properties of the sediments are improved and erosion or dispersion of contaminants is reduced at the same time. Immobilisation of sediments can be done in-situ or ex-situ.

In-situ immobilisation: the SSI-technique

The basic principle of SSI is a combination of the well-known hydraulic jet grouting, and mechanical deep soil mixing methods. Jet grouting involves injection of the soil with grout at very high pressure (VHP grouting) from a rotating lance (a double wall drilling rod). The technique works effectively in sand, which can be liquefied, but is insufficient in clay or sludge, where the soft soil is not liquefied by the jets.

Soil mixing (Deep Chemical Mixing or DCM) is a technique, which mixes in-situ soft soil with a cement grout, using a large-scale multi-shafted agitator. The technique is effective in soft clay, but the required plant is large and expensive because the power needed to mix the clay without pre-cutting is extremely high.

The SSI technique and plant are based on a jet-grouting rig with a special mixing blade on the bottom end of the drilling rod. High-pressure injection nozzles are positioned along the blade. These nozzles help the rotation and penetration of the mixing blade into the soft soil. Low-pressure grout is applied from small diameter openings to fill the void behind the blade as it rotates. Thus SSI combines the advantages of both methods. The jetting process allows the use of a much lighter rig than used for DCM, whilst the rotating blade gives much better mixing of clay or sludge than ordinary jet grouting techniques.

SSI is developed to strengthen and stiffen soft soil, but by adding specific binding agents to the injected cement grout the same technique is suitable to immobilize any possible contaminants present in the treated soft soil or sediments. The technique has already been applied with the purpose of immobilizing heavy metals in river sediments. Some analyses were performed before and after applying the technique in order to evaluate the lixiviation of heavy metals in the river sediments. The results explicitly show the binding of the metals into the sludge, due to the reaction of the particular-binding agents.

In-situ immobilisation: the 3SI-technique

The Surface Sediment Stabilisation by Injection technique is based on the same principle as SSI. Unfortunately, by its concept, SSI works discontinuously by means of overlapping grout columns. When using the 3SI technique, however, the drilling rig is mounted aboard a floating vessel, e.g. a small cutter suction dredger. Combining the rotary injection system with the swinging movement of the cutter suction dredger, allows a perfect three-dimensional positioning of the injection system, and a continuous operation. The mixed layer is solidified or immobilised, and will serve as a capping layer for any polluted sediments underneath. The thickness of the mixed layer can be controlled between 0.5 and 2m.

Both in-situ immobilization techniques have proved successful, when evaluated according to two main criteria. Firstly, the short and long term geotechnical requirements the stabilised sediments should have in order to avoid propeller induced erosion (and hence mechanical transport of pollutants); secondly, the leaching behaviour of the stabilised sediments for the pollutants of interest.

Ex-situ immobilisation: treatment of heavy metals

When sediments are contaminated with heavy metals, they can still be turned into a useful product through immobilization of the pollutants in the material. After the dredging and dewatering processes, the sediments can be mixed with a whole range of additives, depending on the exact pollutant, in order to solidify and immobilise. These additives can be cement-dust, lime, fly ashes, etc.

A technique that has recently been successfully applied is so-called Cold Immobilisation. Cold Immobilisation is based on both physical/mechanical and chemical immobilisation of the contaminants. Physical immobilisation is based primarily on encapsulation of the contaminants and moisture screening. Chemical immobilisation of the contaminants is based on the bond between the contaminants and the additives In respect of inorganic contaminants, above all the physical effect plays a key role, whilst in respect of organic contaminants, the chemical effect is most important.

The manufactured end product is subject to both environmental hygiene-related and civil engineering conditions. It must have physical and mechanical characteristics such that it is suitable for use as (road) foundation or as a surfacing layer. The physical and chemical contaminants in dredged materials are encapsulated and immobilised in such a way, that leaching of contaminating parameters is prevented, and a secondary construction material is manufactured which (even following breaking) still complies with the environmental hygiene requirements laid down in the Construction Materials Bill.

Ex-situ immobilisation: treatment of TBT contamination

Tributyl-Tin (TBT) is an aggressive biocide that has been used in anti-fouling ship paints since the 1970s. The toxicity of TBT prevents growth of algae, barnacles and other marine organisms on the ship's hull. It leaches from the paint, enters the marine environment and accumulates in the sediment, in particular in areas with many ship movements like harbours and ports. TBT is responsible for the disruption of the endocrine system of marine shellfish leading to the development of male sex characteristics in female marine snails. It also impairs the immune system of organisms and shellfish develop malformations after exposure to extremely low levels of TBT in seawater. For these reasons TBT is called "perhaps the most toxic substance ever deliberately introduced to the marine environment by mankind".

The impact of TBT contamination in port sediments on future shipping and port development is significant. Many governments have now developed criteria that can be used to determine whether sediments contaminated with TBT can be disposed of at sea. Different countries take different approaches for setting standards. Belgium, for instance, has set a value of 7µg/kg dw whereas Germany has outlined an approach in which maximum TBT levels will be lowered in accordance with the IMO (International maritime Organisation) rules on the subject. Until 2005, sediments with 600µg/kg dw TBT can be disposed of at sea, whereas this changes to 300µg/kg dw after 2005 and 60µg/kg dw after 2010. If TBT levels in the sediments are above the imposed standards, the dredged sediments cannot be relocated at sea, and have to be taken ashore. This leaves the port authorities with no other choice than dewatering the sediments, and disposing of them in an expensive landfill. Since in many cases this is not an option, this may lead to abandonment of plans to dredge the port or river in question.

DEC as a remediation contractor have encountered this problem at many locations all over Europe, and have invested considerable effort to tackle the TBT issue. At the St. Sampson's Marina project in Guernsey (Channel Islands), DEC successfully dealt with the TBT contamination.

The small harbour of St. Sampson's is being turned into a marina. A sill and gate were to be installed, to keep water in the harbour at low tide. This involved the dredging of some 25,000m³ of sediments, which were contaminated with TBT at concentrations between 640µg/kg and 1770µg/kg. According to local regulations, these levels were too high to allow the sediments to be dumped at sea. Since there are limited options for disposal on the island, the States of Guernsey expressed the intention to clean the sediments and reuse them in a reclamation area off the north side of the harbour.

The biggest concern was to protect the groundwater below the reclamation area from TBT pollution. The sediments in the harbour contained a high percentage of sand and very little organic material, so they were very prone to leaching (average concentrations in leachate were 1,8µg/l). Therefore the remediation level was set to undetectable levels of TBT in the leachate.

The sediment treatment project was awarded to DEC and a first step consisted of laboratory trials to select the additive that could prevent TBT from leaching. Previously, DEC had conducted several feasibility tests on other sediments as well, including the addition of Ordinary Portland Cement or so-called environmental clays.

From these tests it became clear that the addition of cement has dramatic effects on the leaching behaviour of TBT. This can easily be explained when examining the behaviour of TBT at different pH levels, adsorption characteristics of TBT to the different sites available in sediments and parameters that influence this adsorption. It was also noted that the addition of environmental clays had no effect and that the addition of 2%m/m OrganoDEC was sufficient to treat all the sediments of St. Sampson's harbour below the requested detection limit. Since the addition of cement was not advisable due to the leaching behaviour of TBT, it was decided to first dewater the sediments by using the lagooning technique, before treating them and reusing them in the reclamation area.

By the end of 2003, the project was completed successfully. All samples taken of the treated material have shown non-detectable TBT levels in the leachate.

Large-scale treatment centres

As described before, in the last ten years, several cost-effective treatment techniques for dredged material have been developed and implemented on a large scale. Nevertheless one of the biggest problems in the installation of a large-scale treatment centre is finding a suitable and socio-economical acceptable place, especially in harbour areas where space is often occupied or reserved for harbour related industrial use.

Centralising sediment issues in one region, centralised treatment centres share investment costs between different parties and offer economies of scale. When limited or no space is available, a floating treatment centre is an option.

Mobile treatment centres

A modular MobiDEC pontoon is adapted to local needs. It can be fitted with dredging equipment, a dewatering plant, a water treatment installation, etc., according to local needs, legislation and issues. When using a MobiDEC mobile treatment centre, there is no need to transport dredgings by road or rail. This will solve many logistical problems. The pontoon can be shared, sailing from port to port, dredging and treating sediments.

For example, the whole dewatering plant, with a footprint area of 7.5 m wide by 40 m long, can be installed on a floating pontoon. In this way, the dewatering is done in the direct vicinity of the dredging operations or even as part of the dredging. This prevents construction of lagooning fields, long transport of wet sludge, licensing issues, etc.

Beneficial re-use
Once dewatered and stabilised, the sediments can be used for numerous applications. Most obvious is the use as foundation or fill material in the construction industry, e.g. for the building of motorways. Often, the use in capping layers is possible, for landscaping or for covering landfills. Some projects have proven the use for treated sediments as raw material for the production of bricks, artificial gravel, etc.

Conclusion
Dewatering, bioremediation and immobilisation/stabilisation now offer a great opportunity to deal with contaminated sediments, which until recently had to be landfilled. A first priority is the dewatering in multi-layered lagoons, or in mechanical dewatering systems. Once a dry matter content of 65% or higher is reached, bioremediation and soil conditioning techniques, such as immobilisation, are available. Immobilising additives can be used to reduce the spreading of contaminants by leaching, erosion or dispersion.

Innovation and legislative drivers have accelerated the development of several treatment techniques for beneficial re-use of contaminated sediments.

Practical and Effective Environmental Monitoring with reference to Maintenance Dredging for the Port of Maputo, Mozambique

Steve Challinor, Royal Haskoning, Exeter, UK.

Abstract
In this paper, practical and effective environmental monitoring is discussed in the context of the concept of a dredging continuum. Under the dredging continuum concept, the collection and evaluation of dredging information is not conducted in isolation, but forms part of a continuous, cyclical programme of information management throughout the dredging process. This concept is particularly relevant to regular and repeatable maintenance dredging. Accordingly, environmental monitoring should facilitate evaluation, verification, feedback and improvement of the environmental performance of maintenance dredging operation against dredging objectives.

Many factors affect the practicality and efficacy of monitoring maintenance dredging; and many of these many factors need to be considered in a site-specific context to successfully monitor the environmental performance of maintenance dredging. In this paper, the factors affecting practical and effective environmental monitoring are discussed in the context of maintenance dredging of the approach channel to the Port of Maputo in Mozambique, and with particular reference to water quality issues.

Introduction
The Dredging Continuum
Since maintenance dredging is a regular and repeatable operation, the concept of a dredging continuum (IADC & CEDA, 1997; see Figure 1) can be used to illustrate how environmental monitoring needs to be integrated into all stages of the project cycle to make it practical and effective. Conversely, environmental monitoring that fails to integrate itself is at risk of being impractical and ineffective.

Under the dredging continuum, a maintenance dredging project begins with pre-dredge investigations. Concerning environmental monitoring, the pre-dredge investigations may include an environmental impact assessment (EIA) or surveys to determine environmental baseline conditions. As a result of these investigations, the environmental objectives for maintenance dredging are established. Once maintenance dredging is underway, environmental monitoring is undertaken to evaluate the dredging operation against the environmental objectives. As a result of objective evaluation, environmental monitoring provides feedback so that future maintenance dredging operations can be improved. For example, if monitoring identifies environmental impacts that were not predicted, or identifies no environmental impacts when impacts were predicted, then the subsequent pre-dredge investigations, dredging operation, etc may need to be changed to meet revised environmental objectives (IADC & CEDA, 1997).

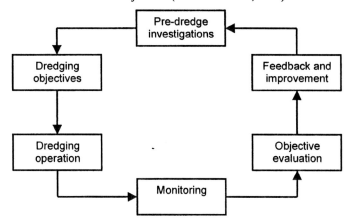

Figure 1 The Dredging Continuum for Maintenance Dredging (adapted from IADC & CEDA, 1997)

Maintenance Dredging for the Port of Maputo

Since 2003, the port of Maputo is operated by a private consortium with a 15 year concession to rehabilitate and develop it into a competitive transit port for import and export markets in southern Africa. Maintaining the port's access channels and berths to -9.4mCD is a priority activity. Under its concession, the consortium is obliged to assess the environmental uncertainties associated with ongoing and future maintenance dredging. The assessment protocol specified environmental monitoring requirements for maintenance dredging. Since the assessment protocol was designed prior to the consortium taking control of the port, the environmental monitoring requirements were not integrated within a dredging continuum. It is in this context that the port of Maputo's maintenance dredging project is used to identify the issues that make monitoring practical and effective.

Pre-dredge Investigations

Environmental monitoring is increasingly required for maintenance dredging. This requirement reflects the proximity of many ports and navigation channels to sensitive environmental locations and the protection the environment is afforded under legislation and policy. For example, the EIA process may require environmental monitoring for reasons such as an evaluation the dredging performance in environmental terms, or the verification of environmental impact predictions.

Pre-dredge investigations are needed to establish the environmental conditions in which the maintenance dredging is to take place. While some information may be available as a result of the EIA process or as published data, it may be necessary to undertake surveys to determine the environmental baselines against which the dredging performance will be monitored. For example, a condition of contract may limit the total suspended solids (TSS) released to the water column as a TSS concentration stated as *'x' percent over the background concentration at 'y' metres from the dredger*. Accordingly, a pre-dredge investigation may be an important step towards identifying and agreeing the boundaries of environmental compliance criteria for a dredging operation.

In the case of maintenance dredging for the port of Maputo, pre-dredge investigations included an EIA process that included baseline surveys to establish sediment type and water quality in terms of turbidity (as TSS), and included numerical modelling to predict environmental impacts associated with the TSS concentration in sediment plumes. The model results predicted that the water quality impact of dredging induced sediment plumes would be very limited.

Despite the EIA's positive prediction, an independent review of the results of the EIA process recommended that further assessment of maintenance dredging was required to address environmental uncertainties. The independent review also established the assessment protocol for the environmental monitoring; stipulating the scope of pre-dredge investigations, including: marine habitat sensitivity mapping, bed sediment characterization for dredging areas, and background turbidity / TSS for near-field areas (within 1000m of dredging) and far-field marine habitats (beyond 5000m of dredging).

Dredging Objectives

Objective setting is crucial to the success of environmental monitoring. IADC & CEDA (1998) identify the following objectives for environmental monitoring: environmental compliance, impact verification, and feedback and improvement.

Objectives concerning environmental compliance of maintenance dredging operations typically focus on water quality impacts and/or ecological impacts. Maintenance dredging may have to comply with a range of environmental compliance criteria that may be legally and/or contractually imposed. Environmental compliance criteria for water quality can include statutory standards or non-statutory standards derived and agreed following pre-dredge investigations. The objective of monitoring is to record whether the dredging causes the environmental criteria to be exceeded.

It is not appropriate to monitor dredging for environmental effects that are unlikely to occur, nor is it practical to monitor for every environmental aspect of dredging on every environmental parameter (John et al, 2000). Therefore, dredging objectives need to be identified that focus monitoring on the site-specific environmental matters that are relevant to the dredging operation to make monitoring practical and effective. For example, where pre-dredge investigations identified that pore water concentrations were highest for arsenic and zinc, monitoring focussed on these metal contaminants only to compare the environmental impact of two maintenance dredging methods at Zeebrugge (Pieters et al, 2002).

Verification of predicted environmental impacts may be an important objective of monitoring. For example, monitoring may provide data that can be used to verify the accuracy of environmental impact prediction based on numerical models. Accordingly, the objectives concerning verification also require focussed monitoring.

Feedback and improvement objectives are necessary to complete the dredging continuum and to help achieve a satisfactory environmental performance during subsequent maintenance dredging operations. These objectives may be related to environmental compliance and verification objectives, as shown by the set of monitoring objectives for maintenance dredging at Maputo, which were:
- to identify the environmental performance of ongoing maintenance dredging based on environmental compliance criteria;
- to verify the findings of a numerical model of ongoing maintenance dredging, based on field data;
- to improve, if necessary, the environmental performance of future maintenance dredging, based on environmental compliance and model verification; and
- to identify long-term environmental monitoring for future maintenance dredging.

Environmental compliance was focussed on the impacts of TSS concentrations in the sediment plumes generated by maintenance dredging. The following environmental compliance criteria were established:
- TSS should be within of 10% of background TSS concentrations in the near-field (within 1000m of the dredging); and
- TSS should not affect marine habitats in the far-field (beyond 5000m of the dredging).

Dredging Operation

Many factors affect the practicality and efficacy of monitoring, including those related to the dredging operation. The maintenance dredging operation undertaken at Maputo can be used to illustrate two of the factors affecting environmental monitoring; these are the type of dredger and its operational characteristics, and the timing of the dredging operation relative to weather conditions.

Until 2003, maintenance dredging at Maputo comprised an approximate annual dredge of 1Mm^3 from the port's navigation channels and berths using a backhoe dredger with a dredging rate of 90m^3 / hour, supported by two barges each having a hopper capacity of 150m^3 (see Figure 2). The maintenance dredging operation changed in 2003 when a trailing suction hopper dredger (TSHD) with a hopper capacity of 5000m^3 (see Figure 3) was used to remove approximately 1.5Mm^3 from a 7.5km stretch of the Polana Channel (the approach channel to the port). The TSHD operated for 24 hours a day for 21 days during December 2003, and was unaffected by the adverse weather conditions and wind-waves characteristic of the wet season. The hopper overflow was not used because significant amounts of sediment settled in the navigation channel within a few hundred metres of the release point.

Figure 2 Backhoe Dredger at Maputo in September 2001 (courtesy of Royal Haskoning; www.royalhaskoning.com) and Figure 3 TSHD at Maputo in December 2003 (courtesy of MER; www.mer.co.za)

In terms of the dredger's characteristics, the use of a TSHD for maintenance dredging at Maputo causes some difficulty for planning consistently representative monitoring stations when environmental compliance is based on TSS concentrations at set distances from a moving dredger: for example, because the sediment release point continuously changes. This factor also makes it difficult to accurately repeat monitoring in subsequent years. For fixed dredgers, such as the backhoe that previously undertook maintenance dredging at Maputo, it is relatively simpler to plan for the monitoring stations to be relocated relative to a repositioned dredger. In retrospect, it is possible that the monitoring protocol for Maputo was designed for a fixed backhoe dredger, rather than a mobile TSHD. Since monitoring protocols are not simply transferable for different dredging methods, this issue serves to emphasise the practical benefits of integrating monitoring into a dredging continuum.

In addition, the TSHD's overflow was not used for most of the maintenance dredge at Maputo because significant amounts of sediment settled rapidly back into the navigation channel. Accordingly, the major release point for a dredging induced sediment plume changed from where the overflow would discharge near the water surface to the vicinity of the suction head at the sea bed. This change of operation a few days into the project meant that the original monitoring protocol would have proved ineffective because it stipulated only monitoring turbidity / TSS at the surface of the water column. Since this was not considered to be best practice, turbidity / TSS monitoring was undertaken at surface, middle and near-bed water depths. This change made the environmental monitoring effective, as can be seen by the results shown in Figure 4.

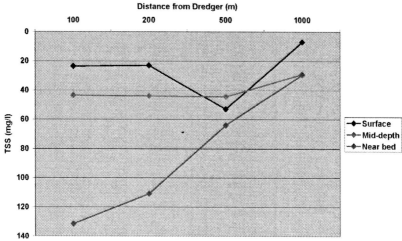

Figure 4 Environmental Monitoring Results for TSS during Dredging at Maputo (Royal Haskoning, 2004b)

In terms of the timing of the dredging operation relative to weather conditions, pre-dredge investigations had identified that Mozambique experiences significant seasonal weather variation which affects background TSS concentrations in Maputo Bay. In the dry season, the wind conditions are calmer and background TSS concentrations are believed to be in the range of 63-319mg/l (Niras Portconsult, 2000). However, maintenance dredging in December 2003 coincided with the wet season, when strong onshore winds cause 1.5m high wind-waves and background TSS concentrations up to 1455mg/l (Royal Haskoning, 2004a). Accordingly, the timing of the dredging operation has significant implications for the practicality and efficacy of monitoring because seasonal and daily weather conditions affect background TSS concentrations. In the case of the environmental objectives set for dredging at Maputo throughout the day during the wet season, the weather variation meant that the limits on TSS concentrations (based on 10% above background at 1000m) varied between 55mg/l and 1600mg/l. There are other examples of how the timing of the dredging operation can coincide with varying environmental parameters, such as seasonal and diurnal fluctuations and one-off storm events affecting TSS concentrations (e.g. Whiteside et al, 1996; Claeys et al, 2001).

Finally, although the TSHD at Maputo could work through the 1.5m high wind-waves, monitoring had to be abandoned during this part of the day for practicality, accuracy, and health and safety reasons.

Monitoring

The success of monitoring requires the use of appropriate monitoring techniques based on the dredging objectives. It is not the intention of this paper to provide a detailed review of the different techniques available to monitor maintenance dredging. However, for reference purposes, Puckette (1998) provides a useful review of sediment plume monitoring techniques, covering in situ water sampling, acoustic monitoring, remote sensing and tracer studies. (Note: Technological developments may have improved the performance and accuracy of monitoring techniques since Puckette was published in 1998.)

In situ water monitoring includes direct real-time measurements using turbidity meters / optical backscatter sensors (OBS), dissolved oxygen meters, conductivity meters, etc, and water sampling for TSS analysis at an offsite laboratory. Acoustic techniques can also be used to measure backscatter intensity as the acoustic energy reflected by particulate matter in the water column, although calibration may be difficult. Remote sensing techniques such as aerial photography and satellite imagery may be limited by weather and daylight, and are generally not effective for monitoring sediment plumes. Particulate tracer dyes may be useful in some situations, such as identifying the long-term movement of sediment plumes (Puckette, 1998).

For the purposes of monitoring the maintenance dredging for the port of Maputo, the monitoring protocol established that OBS measurements and water samples were to be taken to produce turbidity data and TSS data sets respectively. The baseline turbidity and TSS data sets made during pre-dredge investigations were correlated to allow the monitoring method (i.e. OBS measurements) to give real-time indications of TSS concentrations during the maintenance dredging. It is necessary to have knowledge about the local correlation of turbidity and TSS because there is no universal correlation that can be used confidently (Thackston and Palermo, 2000).

Given the inflexibility of the monitoring protocol, the OBS correlated TSS concentrations provided the best monitoring method with regard to the dredging objectives set for the maintenance dredging at Maputo. The data could be compared to environmental compliance data and facilitated numerical model verification.

The simplicity of OBS measurement is one of the practical benefits of this monitoring technique. In some places, simplicity can be important consideration for environmental monitoring because practical issues, such as the robustness of the equipment, availability of spares, and ease of maintenance and repair, may outweigh the technical advantages of more complex equipment. In addition, a relatively simple technique may also be the most cost-effective technique.

Evaluation of Objectives

Regarding the environmental compliance objectives, continuous evaluations of a dredging operation can take place in real time. For example, OBS monitoring data may be transferred (via a cable or telemetric link), calibrated and assessed immediately, or fluorescent tracer particles in a sediment plume may be tracked by laser detection (Marsh, 1994). Alternatively, evaluations may occur retrospectively if sample / data retrieval requires subsequent analysis / calibration before evaluations may be undertaken. For example, retrospective evaluation will be necessary if OBS measurements need to be calibrated against TSS samples, acoustic data need to be calibrated and processed through a software package, aerial photographs need processing and analysis, and/or tracer dye particle tracking needs sampling and analysis.

In the case of maintenance dredging at Maputo, the evaluation of environmental compliance was undertaken retrospectively because real time assessment was not necessary under the monitoring protocol's objectives. The monitoring recorded surface, mid-depth and near bed TSS concentrations at 7mg/l, 29mg/l and 29.5mg/l respectively at 1000m from the maintenance dredging operation. A simple evaluation against the pre-dredge background TSS data indicates that the sediment plume's TSS concentrations meet the environmental compliance objective for the near-field zone during calm weather, but become indistinguishable from naturally high TSS concentrations during the daily storm events that occur during Mozambique's wet season.

Verification objectives are likely to be evaluated retrospectively, using the monitoring data to test predictions made about the dredging operation and its environmental impacts. For example, a numerical model may need to be re-run having been calibrated with the monitoring data to verify whether its predictions were correct.

OBS measurements were used to evaluate the numerical model predictions for TSS in sediment plumes generated during maintenance dredging at Maputo. The verification process was undertaken retrospectively because the model had to be re-run to take into account differences between the modelled and actual maintenance dredging; for example, the change to the dredger's sediment release rate when the overflow was turned off. The verification objective evaluation indicated that the model predictions of TSS exceed monitored TSS in the near-field, and that the sediment plume may extend into the far-field on some tides (see Figure 5).

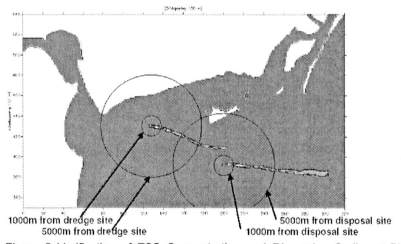

Figure 5 Verification of TSS Concentration and Dispersion Sediment Plumes from Maintenance Dredging at Maputo (Royal Haskoning, 2004b)

Feedback and Improvement

Feedback and improvement is the final stage of each cycle of the dredging continuum. In addition to monitoring for environmental compliance and verification objectives, monitoring will be more effective if it is fed into and used to improve the environmental performance of subsequent maintenance dredging operations.

In the case of dredging for the Oresund Link, a feedback monitoring programme was put in place to maintain approval for dredging operations and to through compliance with environmental criteria. For this project, monitoring was integrated with numerical modelling to facilitate re-evaluation of predicted impacts and provide feedback to the dredging project within the terms of the contract (Jensen and Lyngby, 1999).

In the case of maintenance dredging at Maputo, the feedback and improvement process identified that the practicality and efficacy of environmental monitoring was limited by restricted data acquisition over a short-term monitoring period. Also, it was identified that adverse daily weather conditions posed health and safety risks for sampling from a small vessel, while large natural increases in TSS concentrations (up to 1455mg/l) caused the sediment plume from the dredging operation to become indistinguishable from background sediment loads within a few hundred metres of the dredger.

Conclusions

The example of environmental monitoring of maintenance dredging for the port of Maputo has illustrated that there are many factors that affect the practicality and efficacy of monitoring, including the type of dredger, dredging operations, monitoring objectives, technical complexity of monitoring equipment, environmental variability, working conditions, background data sets, and costs. However, by using the dredging continuum concept, practical and effective environmental monitoring is applied at all stages of the maintenance dredging process, and is facilitated towards meeting its objectives.

References

Claeys, S., Dumon, G., Lanckneus, J. and Trouw, K. (2001). Mobile turbidity measurement as a tool for determining future volumes of dredging material in access channels to estuarine ports. Terra et Aqua, No. 84, September 2001, pp.8-16.

IADC & CEDA (1997). Environmental aspects of dredging. Report 3. Investigation, interpretation and impact. pp.67.

IADC & CEDA (1998). Environmental aspects of dredging. Report 4. Machines, methods and mitigation. pp.80.

Jensen, A. and Lyngby, J. E. (1999). Environmental management and monitoring at the Oresund Fixed Link. Terra et Aqua, No. 74, March 1999, pp.10-20.

John, S. A., Challinor, S. L., Simpson, M., Burt, T. N., and Spearman, J. (2000). Scoping the assessment of sediment plumes from dredging. CIRIA Report C547. Construction Industry Research and Information Association, London, pp.188.

Marsh, J. (1994). Monitoring of a dredging operation using innovative sediment tracing techniques. Dredging '94, Volume 2, pp1360-1369, Proceedings of the Second International Conference on Dredging and Dredged Material Placement. ASCE.

Niras Portconsult (2000). Maputo port privatisation and rehabilitation project, Mozambique. Environmental impact assessment. Prepared on behalf of MPDC.

Pieters, A., Van Parys, M., Dumon, G. and Speleers, L. (2002). Chemical monitoring of maintenance dredging operations at Zeebrugge. Terra et Aqua, No. 86, March 2002, pp.3-10.

Pukette, T. P. (1998). Evaluation of dredged material plumes. Physical monitoring techniques. DOER Technical Notes Collection (TN DOER-E5), US Army Engineer Research and Development Center, Vicksburg, MS.

Royal Haskoning (2004a). Maputo port dredging assessment. Report 1 Baseline environmental conditions in Maputo Bay. Sectional Report. Prepared on behalf of MPDC.

Royal Haskoning (2004b). Maputo port dredging assessment. Report 4. Monitoring and verification. Sectional Report. Prepared on behalf of MPDC.

Thackston, E.L. and Palermo, M.R. (2000). Improved methods for correlating turbidity and suspended solids for monitoring. DOER technical Notes Collection (ERDC TN-DOER-E8), US Army Engineer Research and Development Center, Vicksburg, MS.

Whiteside, P., Ng, K. and Lee, W. (1996). Management of contaminated mud in Hong Kong. Terra et Aqua, No. 65, December 1996, pp.10-17.

Discussion

Regulation
Concern was expressed that unnecessary, or conflicting regulations, could result from the interpretation of various EU Directives, in particular the Water Framework and Habitats Directive. Government representatives expressed optimism, but this optimism was not shared by all speaker and delagates.

Water Framework Directive
It was noted that for many water bodies, the '*status quo*' is not be easily defined.

Disappointment was expressed over the apparent decision by DEFRA that maintenance dredging should be classified as a 'Plan' or 'Project'. This will bring it outside of the scope of Management Plans and greatly increase the administrative load attached to maintenance dredging. It was suggested that strictly, a decision to cease regular dredging could amount to a Plan or Project, which is unlikely to have been the intention when the Directive was drafted.

It was agreed that it is important that close and regular dialogue between the industry and regulators be maintained and that regulators should adopt a constructive and helpful approach.

In response to a question, is a dredged channel a 'modified water'? The opinion given was YES.

It was noted that within the Water Framework Directive, sediment is illogically classified as a contaminant, apparently in ignorance of the fact that in most European waters sediment is a natural ingredient in the fluvial and marine environments. It was reported that 'Sednet' is addressing this matter.

It was asked, is 'Trickle charging' likely to be banned as a consequence of the WFD? In response it was considered that the possibility does exist, the root problem being that even clean sediment is classified by WFD as a contaminant.

It was noted that there are areas in which there is direct conflict between the WFD and the Habitats Directive.

In response to a question, what does 'artificial water body' mean, it was stated that it is defined as being, '*where no water previously existed*'. Some water bodies could be classified as part modified and part not.

It was stated that ports should be aware of the benefits of being classified as a '*modified water body*'.

Maintenance Dredging II, Thomas Telford, London, 2005.

Environmental Liability
It was recognised that the effect of the Environmental Liability Directive is likely to be increased cost. How is '*significant impact*' to be assessed and by whom? The liability of operators may be protected if using '*state of the art methods*'.

Landfill Directive
It was noted that on the River Thames the Cliffe Disposal site receives liquid waste, which is in conflict with the Landfill Directive. It was thought by various speakers that the Regulator generally takes a pragmatic view of such cases.

Should dredged material be classified as Waste? Definitely not, was the general view and it was noted that classification as waste is not consistent across Europe, which serves to underline the problem of different interpretations being placed on European Directives when transposed into national law.

General
In relation to 'beneficial use' it was reported that DEFRA has recently expressed the view that an increase of 10% in cost to achieve beneficial use should be seen as reasonable.

It was agreed that experience has shown that Regulators are inclined towards a precautionary approach, which, whilst satisfactory in principle, causes difficulty if excessive caution is applied.

Arising from the discussions on regulation, it was generally agreed that the Dibden Bay situation, wherein a figure of £45 million is quoted as having been expended by ABP in studies for planning and Public Inquiry, only to result in planning refusal, cannot be justified. There was broad agreement that the procedures of project assessment and planning must be simplified to reduce time and cost to a more appropriate level.

It was agreed that all parties with an interest in dredging in the UK and in Europe generally, should make strenuous efforts to become involved in the consultation process, so as to minimise the risk of unworkable or unrealistic regulation being transposed into law. All interested parties need to speak with a common voice. Data needs to be collected and check lists prepared to ensure compliance with regulations.

The practicalities of doing it
Hydrodynamic and non-dredging solutions
It was reported that experience in the Port of Hamburg indicates that water injection may not be effective in areas of fluid mud, or in areas of flat seabed.

Recent developments in Trailer Suction dredgers
In relation to the installation of horizontal diffuser plates in the hopper to improve settling, it was asked if these tend to become ineffective due to blockage by debris? It was reported that experience to date has shown this to be only a minor problem.

Inland projects
It was noted that dredged material arising from Barton Broad was disposed of on agricultural land and it was asked if the landowner was compensated? In response it was stated that compensation for loss of use during period of dredging and prior to restoration was given,

but no further compensation was required, because the post-restoration productivity of the land was not adversely affected.

Contract management
Forms of Contract
Concerning 'Environmental liability' it was asked, what limit is there on contractors liability. After some discussion it was apparent that the legal position is not yet clear, but it was considered by some that liability should not exceed the contract value.

In response to a question on how to avoid conflict, it was stated that the risk of conflict is minimised if the task to be performed and the responsibilities of the parties to the contract are very clearly defined from the beginning.

Contract v 'In-house' dredging
It was reported that experience has shown that operators of direct labour dredging operations often fail to adopt realistic accounting when comparing in-house costs with those of contract dredging. This can result in prolonged uneconomic operation.

Environmental aspects
Prediction and modelling transport & sedimentation
It was asked, how effective is the use of ADCP methods when measuring suspended solids at the bottom of the water column? In reply it was stated that the method is not able to measure concentration at the bottom of the water column.

Planning environmental protection
In relation to use maintenance dredging on River Thames, it was asked, what proportion of work was by water injection methods? In response it was stated that approximately 40% to 50% is by water injection, but this percentage applies only to the lower river reaches and the method is not used where sediments contain significant levels of contaminants.

In relation to the Port of London's geographical database approach to environmental protection, it was asked if other methods had been considered? In reply it was stated that other methods have been considered and systems employed at other North Sea locations, in particular in Holland, are under review.

It was reported that at Manchester Ship Canal, the flood protection function of the canal and its maintenance are an important consideration. A delegate expressed the view that past fluvial flooding of the Upper Thames was the result of too little maintenance dredging.

Printed in the United Kingdom
by Lightning Source UK Ltd.
109793UKS00001BA/75